Nonlinear Dimensionality Reduction Techniques

Sylvain Lespinats • Benoit Colange • Denys Dutykh

Nonlinear Dimensionality Reduction Techniques

A Data Structure Preservation Approach

 Springer

Sylvain Lespinats
National Institute of Solar Energy (INES)
Grenoble Alpes University
LE BOURGET-DU-LAC, France

Benoit Colange
National Institute of Solar Energy (INES)
Grenoble Alpes University
LE BOURGET-DU-LAC, France

Denys Dutykh
CNRS - LAMA UMR 5127
Université Grenoble Alpes
Université Savoie Mont Blanc
Campus Scientifique
F-73376 Le Bourget-du-Lac, France

ISBN 978-3-030-81028-3   ISBN 978-3-030-81026-9   (eBook)
https://doi.org/10.1007/978-3-030-81026-9

This Springer imprint is published by the registered company Springer Nature Switzerland AG
The registered company address is: Gewerbestrasse 11, 6330 Cham, Switzerland

To everyone who introduced me to the fun of science including my parents, Mr. Fouillard, Mr. Laurent and Pr. Fertil
(SL)

To my friends and family
(BC)

To Katya, Nicolas and Michel
(DD)

Foreword

It is my honor to write this foreword. I had the chance to work with Drs. Benoît Colange, Sylvain Lespinats, Jakko Peltonen, and Denys Dutykh on ClassNeRV, an original and efficient way to compute supervised dimensionality reduction (DR). I have known Sylvain for years as we explored different aspects of DR interpretation (CheckViz) and proposed an early version of ClassNeRV called ClassiMap. In short, we designed a stress function that directs distortions where they have the least impact on the class structure. We wanted to translate this approach into the probabilistic framework of stochastic neighbor embedding techniques (tSNE, NeRV) with the help of Jaakko, one of NeRV's authors. Jaakko was instrumental to interpret the equations of ClassiMap in the new framework, but we got stuck with the constraints of the Kulback–Leibler-based formulation of the stress function. Fortunately, Sylvain and Denys engaged Benoît for a PhD to develop new tools to analyze energy-related data. Benoît solved our problem in no time, coming up with a very elegant and unexpected solution. I let you discover it in Chap. 6 of this book which outlines the core of Benoît's PhD project. Long story short, it is my real pleasure to introduce this work.

At a time where we face challenging health, financial, and environmental crises, data are collected, stored, and processed at a larger and larger scale in a tentative to better model these phenomena and tackle these problems. Alas, the human brain did not scale up at the same pace, and we bear the same cognitive power as our millennial ancestors. The consequence is these data need to be drastically summarized for us to understand.

Mathematicians defined many summary statistics, from basic numerical values like quantiles to more advanced statistical models (nonlinear regression, deep neural networks, generative, adversarial, and causal models) and more recently topological ones based on Morse–Smale theory and persistent homology, etc. Despite this undoubtful progress on the modelization side, end-users of these approaches still trust more their naked eyes than a few or even many numerical indicators to detect patterns these models could have missed. And they are right, some patterns escape detection by the most advanced of the models, or are detected but not explicitly encoded, pushing the need for more interpretable and even explainable models that let the user know what patterns the model discovered. In short, we will not avoid the need for data visualization and visual analytics, and multidimensional projections and related techniques are among these approaches which can preserve quite a broad range of similarity-based patterns difficult to detect with more automatic models.

I have grown my expertise for 20 years in both these fields, machine learning for clustering, dimension reduction, and topological data analysis on the one hand, and visual analytics for guiding the interpretation of these models, on the other hand, seeking techniques that would bridge the gap between high-dimensional data and human understanding. Therefore, I am well-versed in these domains to appreciate the depth and breadth of the present work and its foreseeable impact.

By reading this book you will touch upon some of the toughest challenges faced by data scientists and domain experts when it comes to explore and analyze high-dimensional continuous data in search of meaningful patterns. The authors propose original approaches to estimate local data intrinsic dimensionality (TIDLE); to compute meaningful maps that account for that information (ASKI) and better preserve the structure of the classes (ClassNeRV, ClassJSE); to help interpret the visual patterns these maps display (MING) avoiding misuse or disuse of the map; to evaluate maps' quality with new class-aware indicators; and to map out-of-sample data onto an existing layout. Technical details and context are given with all the necessary explanations for non-experts to enrich their skill set. The proposed techniques are applied to various analytic problems in smart-building, photovoltaics, and batteries that will inspire the practitioner to try out the same in their own domain. At last, students and researchers will find an excellent introduction to the topic and food for thought to further their own research projects. Finally, I hope you will enjoy the reading as I did.

Doha, Qatar   Dr. Hab. Michaël Aupetit
May 24, 2021   Senior Research Scientist at
                                                  Qatar Computing Research Institute
                                                    Hamad Bin Khalifa University
                                                         maupetit@hbku.edu.qa
               https://www.hbku.edu.qa/en/staff/dr-michael-aupetit/

Preface

A teacher can never truly teach unless he is still learning himself. A lamp can never light another lamp unless it continues to burn its own flame. The teacher who has come to the end of his subject, who has no living traffic with his knowledge but merely repeats his lessons to his students, can only load their minds; he cannot quicken them. minds; he cannot quicken them.

Rabindranath Tagore

Performing diagnosis of an energy system requires to identify relations between observed monitoring variables and the associated internal state of the system. Dimensionality reduction, which provides a visual summary of a multidimensional dataset, constitutes a promising tool to help domain experts to explore these relations. This book reviews existing techniques for visual data exploration and dimensionality reduction, proposes new solutions to challenges in that field and illustrates them on several practical cases.

Monitoring variables may be, for instance, I–V curves (i.e. current versus voltage characteristics) for photovoltaic systems or acoustic signals for batteries. The values of monitoring variables associated with a given operating state of the system, either normal or abnormal (i.e. a fault), constitute the signature of that state. These signatures may be treated as multidimensional data, that is data characterized by many variables, or more generally as metric data. Indeed, most of such methods may also be used when data are only known by distances between them. Visual exploration consists in exploiting dimensionality reduction (DR) methods to display proximity relations in the dataset as a 2D or 3D scatterplot, also referred to as a *map*.

In such a map, each data is represented by a point positioned so that the Euclidean distances between points reflect as faithfully as possible the dissimilarities between the associated data (in the sense of the metric chosen for the data). Therefore, the map enables to forge an intuition on the data structure, discovering groups of similar points. It also allows to form hypotheses about the relations between the considered data and auxiliary variables which may be displayed on the map. In the context of diagnosis, a map may, for instance, be used to identify unknown operation states, by locating groups of similar signatures. It may also help to study the link between signatures and known states of the system. Thereby, it enables to assess the discriminability of different faults based on their signatures. This discriminability property is tied to the considered monitoring variables and is a necessary condition for diagnosis. Finally, once a map of signatures associated to discriminable operation states is obtained, the different regions of that map may be characteristic of specific states. Hence, new signatures associated to unknown states could be positioned a posteriori on this map in order to diagnose the state of the system.

Several dimensionality reduction methods focus on neighbourhood information. It means that the neighbours of a point in the map, in terms of the Euclidean distance, should correspond as faithfully as possible to its neighbours in the data space. Following this paradigm, two types of distortions may be identified: false neighbours, which are points represented too close from each other in the map, and missed neighbours, which are represented too far from each other.

However, the construction of a map is subject to many challenges. Indeed, the data structure is often too complex to be perfectly represented in two or three dimensions. Two types of distortions can be distinguished: false neighbours, which correspond to pairs of points represented too close from each other in the map, and missed neighbours, represented too far. One of the main differences between DR technics lies in the distortion penalization that balances their relative criticality. Once a map is obtained, the overall level of distortion of the data structure can be assessed with scalar indicators. However, because these indicators account for a subjective choice, finer methods for maps evaluation are also possible. Such local evaluation tools are needed for a fair analysis of maps in order to prevent biased inferences induced by distortions.

In order to provide a map with the least distortion of relevant information, this book proposes supervised dimensionality reduction techniques, called **ClassNeRV** and **ClassJSE**, which benefit from available class information. They preserve the neighbourhood structure, while steering unavoidable distortions to reduce their impact on the representation of classes. This amounts to mainly avoiding missed neighbours within a same class and false neighbours between different classes. An unsupervised technique, named **ASKI**, is also presented. It automatically adapts to the effective dimensionality discrepancy between the data and the map, hence preserving neighbourhoods, while not excessively distorting distances between points if possible. Once a map is obtained, the analyst must be aware of the unreliable information that it may convey. To meet this need, class-aware quality indicators enable to assess the overall preservation of classes structure by the map. Moreover, **MING** helps to locate distortions, thus supporting a less biased analysis of the map. It presents with coloured edges, the reliable neighbours of each point, as well as its false and missed neighbours. Other challenges in the analysis of multidimensional data are addressed. To assess the complexity of a dataset, **TIDLE** provides a local estimation of its intrinsic dimensionality, while the radial basis function (RBF) interpolation can be applied for mapping-agnostic out of sample extension. These methods are then applied to the representation of expert-designed fault indicators for smart-buildings, I–V curves for photovoltaic systems and acoustic signals for Li-ion batteries. It allows to discover co-occurrences of faults, differences of behaviours between power plants and relations between measured signals and variables of interest.

The present book and the results presented here come from the PhD thesis of the second author (BC). His work was advised and supervised by the two other authors (SL and DD). In particular, SL and DD would like to take advantage of this occasion to thank Benoît for joining our team for three nice and productive years which greatly boosted our cooperation. We wish you all the best for your bright professional future.

Aix-les-Bains, France   Sylvain Lespinats

Chambéry, France   Benoît Colange

Chambéry, France   Denys Dutykh
May 2021

Acknowledgements

We would like to thank Dr. Michaël J. Aupetit (Qatar Computing Research Institute, HBKU, Qatar) and Prof. Jaakko Peltonen (Tampere University, Finland) for helpful discussions on the DR matters, as well as Prof. Laurent Vuillon (Savoie Mont-Blanc University, France) for his insights in graph theory. Without you this book would not look the same.

We also would like also to thank our colleagues with whom we collaborated, especially, on the applications of DR methods:

- Dr. Nicolas Guillet, Florence Degret and Dr. Vincent Gau for our common work on acoustic signals for batteries
- Alexandre Plissonnier and Arnaud Revel on photovoltaic I–V curves
- Dr. Amal Chabli and Dr. François Bertin on diagnostic on photovoltaic cells
- Dr. Hugo Geoffroy and Dr. Julien Berger on building performances monitoring;
- Dr. Olivier Bastien on various subjects

Funding

This work has be supported by the CEA-INES through the Contract of Formation through Research of Benoît Colange, as well as by the French National Research Agency, through Investments for Future Program (ref. $ANR-18-EURE-0016-$ Solar Academy). We would also like to acknowledge the "Programme d'Investissements d'Avenir—INES.2S" (Grant $ANR-10-IEED-0014-01$) for its generous financial support.

Contents

Acronyms

1D	One dimensional
2D	Two dimensional
3D	Three dimensional
ADEB	Attribute-Driven Edge Bundling
AE	Acoustic Emission
ASKI	Adaptive Student Kernel Imbedding
BIC	Bayesian Information Criterion
BFGS	Broyden–Fletcher–Goldfarb–Shanno algorithm
catSNE	class-aware t-distributed Stochastic Neighbour Embedding
CCA	Curvilinear Component Analysis
CDA	Curvilinear Distance Analysis
CDF	Cumulative Distribution Function
CIE	Comission Internationale de l'Éclairage
CGNE	Class-Guided Neighbourhood Embedding
ClassNeRV	Class-guided Neighbourhood Retrieval Visualizer
ClassJSE	Class-guided Jensen Shannon Embedding
CLT	Central Limit Theorem
cMDS	Classical Multi-dimensional Scaling
COIL	Columbia Object Image Library
DBSCAN	Density-Based Spatial Clustering of Applications with Noise
DD-HDS	Data-Driven High Dimensional Scaling
DR	Dimensionality Reduction
DTW	Dynamic Time Warping
EE	Elastic Embedding
EM	Expectation Maximization
EMST	Extended Minimum Spanning Tree
ES-LLE	Enhanced Supervised Locally Linear Embedding
FDA	Fisher Discriminant Analysis
FDEB	Force-Directed Edge Bundling
FLLE	Fisher Locally Linear Embedding
GGG	Generative Gaussian Graph

GPU	Graphical Processing Unit
HDBSCAN	Hierarchical Density-Based Spatial Clustering of Applications with Noise
Hidalgo	Heterogeneous Intrinsic Dimension Algorithm
HSV	Hue Saturation Value
HSSNE	Heavy-tailed Symmetric SNE
i.i.d.	Independent and Identically Distributed
itSNE	Intrinsic tSNE
IPA	International Phonetic Alphabet
Isomap	Isometric feature mapping
itSNE	Intrinsic t-distributed Stochastic Neighbour Embedding
JSE	Jensen Shannon Embedding
KDA	Kernel Discriminant Analysis
KDE	Kernel Density Estimation
KDEEB	Kernel Density Estimation-based Edge Bundling
KL	Kullback–Leibler
KPCA	Kernel Principal Component Analysis
LAMP	Local Affine Multidimensional Projection
L-BFGS	Low-memory Broyden–Fletcher–Goldfarb–Shanno algorithm
LCMC	Local Continuity Meta Criterion
LDA	Linear Discriminant Analysis
LFDA	Local Fisher Discriminant Analysis
LE	Laplacian Eigenmaps
LLE	Locally Linear Embedding
LMDS	Local Multi-dimensional Scaling
LOF	Local Outlier Factor
MDS	Multi-dimensional Scaling
MNIST	Mixed National Institute of Standards and Technology
MRRE	Mean Relative Rank Error
MST	Minimum Spanning Tree
NCA	Neighbourhood Component Analysis
NE	Neighbourhood Embedding
NeRV	Neighbourhood Retrieval Visualizer
NMF	Non-negative Matrix Factorization
NN	Nearest Neighbours
PAM	Partitioning Around Medoids
PCA	Principal Component Analysis
PDF	Probability Density Function
PHATE	Potential of Heat-diffusion for Affinity-based Transition Embedding
PLLE	Probability-based Locally Linear Embedding
PSD	Power Spectral Density
PV	Photo-Voltaic
RBF	Radial Basis Function
RF-PHATE	Random-Forest Potential of Heat-diffusion for Affinity-based Transition Embedding

RGB	Red Green Blue
S-XXX	Supervised XXX
SGD	Stochastic Gradient Descent
SNE	Stochastic Neighbour Embedding
SNLM	Sammon Non-linear Mapping
SoC	State of Charge
SPCA	Sparse Principal Component Analysis
SPLOM	Scatter PLOt Matrices
SVD	Singular Value Decomposition
SVM	Support Vector Machine
TIDLE	Two-nearest neighbours Intrinsic Dimensionality Local Estimator
tSNE	t-distributed Stochastic Neighbour Embedding
ttSNE	twice Student Stochastic Neighbour Embedding
UMAP	Uniform Manifold Approximation and Projection
UsC	Ultrasound Characterization
YlGnBu	Yellow Green Blue
YlOrRd	Yellow Orange Red

Nomenclature

In the context of Dimensionality Reduction, when a variable as a dual definition in the data and embedding spaces, it is most often denoted by a Greek letter in the data space and a corresponding Latin letter in the embedding space.

Metric Data

$\mathcal{D}, \mathcal{E}, \mathcal{M}, \mathcal{K}$	Data space, embedding space, data manifold, kernel expansion space
δ, d, ∂	Dimensionality of the data space, embedding space and data manifold
i, j	Points indices
ξ_i, x_i	Point i in the data or embedding space
Δ_{ij}, D_{ij}	Distance between point i and j in the data or embedding space
γ_{ij}, g_{ij}	Similarity between point i and j in the data and embedding space
ρ_{ij}, r_{ij}	Neighbourhood rank of point j in the neighbourhood of point i in the data and embedding space

Dimensionality Reduction

τ	Trade-off parameter
w_{ij}	Weight associated to neighbourhood relation $i \sim j$
ζ	Stress function
Φ, Ψ	Mappings between high and low-dimensional spaces
$\breve{\mathscr{I}}, \mathscr{I}$	Quality and distortion indicators for map evaluation
$\mathscr{F}, \mathscr{M}$	False and missed neighbour distortion indicators

Neighbourhood Characterization

$i \sim j$	Neighbourhood relation between point i and j
κ	Number of neighbours
σ_i, s_i	Scaling factor associated to point i in the data and embedding space
λ_i, l_i	Degree of freedom of a kernel associated to point i in the data and embedding space
β_{ij}, b_{ij}	Neighbourhood membership degree of point j in the neighbourhood of point i in the data and embedding space
$\mathscr{D}_{\text{KL}}, \mathscr{D}_{\text{gKL}}$	Classical and generalized Kullback–Leibler divergences

Class-Information

L_i	Class-label of point i
q_{ij}	Class-community of point i and j
$\mathcal{S}_i^{\in}, \mathcal{S}_i^{\notin}$	The set of intraclass and interclass points for point i

List of Figures

List of Tables

Chapter 1
Data Science Context

[The answer of a fully unsupervised Artificial Intelligence to]
the Great Question of Life, the Universe and Everything [is] 42.
The Hitchhiker's Guide to the Galaxy
Douglas Adams

This chapter positions Dimensionality Reduction (DR) in the broader context of data science, considering both its use as an automated pre-processing tool extracting variable (manifold learning) for other automated tasks (e.g., classification, clustering, regression), and as a mapping technique allowing to visualize multidimensional data in a low-dimensional space (spatialization). In the process, it also introduces some general tools of data analysis such as distances computations and intrinsic dimensionality estimation, which are also of paramount importance when performing the Dimensionality Reduction .

First, Sect. 1.1 describes several types of data, showing that they may all be handled as elements of a metric space, which is the common input of most DR algorithms. Then, Sect. 1.2 presents automated tasks of descriptive and predictive data analyses, while defining the data structures that may be discovered using the algorithms designed for those tasks (e.g., clusters, outliers, hierarchies, manifold). Section 2.2 dives into the practical estimation of the intrinsic dimensionality of a dataset, namely the number of degrees of freedom required to fully describe that dataset. Finally, Sect. 1.3 explains how to integrate human cognition in the process of analysing data through visual exploration, which relies on visual encoding of multidimensional data. In that regard, spatialization combined with scatter plots representations constitutes an encoding of those data.

1.1 Data in a Metric Space

This section presents how different types of data (multidimensional data, sequence data, network data) may be characterized in terms of dissimilarities or similarities measured among all pairs of data instances. This amounts to considering all these

S. Lespinats et al., *Nonlinear Dimensionality Reduction Techniques*,
https://doi.org/10.1007/978-3-030-81026-9_1

data instances as elements of a metric (or pseudo-metric) space, or of an inner product space.

1.1.1 Measuring Dissimilarities and Similarities

Data instances of any type may be considered as points in a metric space as long as one may define a metric or distance function to measure the dissimilarity between two instances. This metric space $(\mathcal{D}, \Delta)$ is a topological space equipped with a distance Δ, which provides for each pair of elements of that space a numerical score of their dissimilarity. This proper notion of distance is defined by:

Definition 1.1 A function $\Delta : \mathcal{D} \times \mathcal{D} \to \mathbb{R}^+$ is a distance (or metric) over the space $\mathcal{D}$ if and only if it satisfies the following conditions for all $\xi_i, \xi_j, \xi_k \in \mathcal{D}$:

- $\Delta(\xi_i, \xi_j) \geqslant 0$ (non-negativity),
- $\Delta(\xi_i, \xi_j) = 0$ iff $\xi_i = \xi_j$ (identity of indiscernibles),
- $\Delta(\xi_i, \xi_j) = \Delta(\xi_j, \xi_i)$ (symmetry),
- $\Delta(\xi_i, \xi_j) \leqslant \Delta(\xi_i, \xi_k) + \Delta(\xi_k, \xi_j)$ (triangle inequality or sub-additivity).

Those distances extend to abstract spaces the spatial notion of distance in our three-dimensional physical space, measured using the Euclidean distance (see Sect. 1.1.4). As a tool for measuring dissimilarities, one may also consider pseudo-metrics which do not satisfy all properties of the Definition 1.1. When not otherwise stated dissimilarities between data are computed with the Euclidean distance.

Metric spaces are a more general case of normed vector spaces, that is spaces equipped with a norm $|| \cdot ||$ measuring the size of a vector, defined as follows:

Definition 1.2 A function $|| \cdot || : \mathcal{D} \longrightarrow \mathbb{R}^+$ is a norm if and only if it satisfies the properties for all $\xi_i, \xi_j \in \mathcal{D}$ and $\alpha \in \mathbb{R}$:

- $||\alpha \xi_i|| = |\alpha| \, ||\xi_i||$ (homogeneity),
- $||\xi_i|| = 0 \Rightarrow \xi_i = 0$ (separation),
- $||\xi_i + \xi_j|| \leqslant ||\xi_i|| + ||\xi_j||$ (triangle inequality).

In a normed vector space, a distance is naturally defined between all pairs of point by computing the norm of their difference:

$$\Delta(\xi_i, \xi_j) = ||\xi_i - \xi_j||. \tag{1.1}$$

Normed vector spaces include the subcase of inner product spaces (equipped with an inner product $\langle \cdot, \cdot \rangle$). An inner product must satisfy the following definition:

Definition 1.3 A inner product $\langle \cdot, \cdot \rangle : \mathcal{D} \times \mathcal{D} \longrightarrow \mathbb{R}$ is a positive-definite symmetric bilinear form, namely for all $\xi_i, \xi_j \in \mathcal{D}$ and $\alpha_i, \alpha_j \in \mathbb{R}$:

- $\langle \xi_i, \xi_j \rangle = \langle \xi_j, \xi_i \rangle$ (symmetry),
- $\langle \alpha_i \xi_i + \alpha_j \xi_j, \xi_k \rangle = \alpha_i \langle \xi_i, \xi_k \rangle + \alpha_j \langle \xi_j, \xi_k \rangle$ (linearity),
- $\langle \xi_i, \xi_i \rangle \geqslant 0$ and $\langle \xi_i, \xi_i \rangle = 0 \Rightarrow \xi_i = 0$ (positive-definiteness).

Inner products provide some measure of similarity between two vectors, but it also naturally defines the following norm (which in turn leads to a measure of distance using Eq. (1.1)):

$$\|\xi_i - \xi_j\| = \sqrt{\langle \xi_i - \xi_j, \xi_i - \xi_j \rangle}.$$

A common representation of a metric dataset is the *distance matrix* $\mathbf{\Delta}$. Its element Δ_{ij} indicates the distance between the ith and jth points of the dataset. Due to the properties defining a distance, this matrix is necessarily symmetric and all its diagonal elements are zero. Indeed, symmetry and identity of indiscernibles respectively imply $\Delta_{ij} = \Delta_{ji}$ and $\Delta_{ii} = 0$. This notation can be extended to the general case of dissimilarity matrices.

Alternately datasets in an inner-product space may be represented by their *Gram matrix* $\mathbf{\Gamma}$ whose elements γ_{ij} are the inner products $\gamma_{ij} = \langle \xi_i, \xi_j \rangle$. This extends to the more general case of similarity matrices.

Conversely, any given distance matrix may be interpreted as the one computed from a set of data points in an Euclidean space. The existence of such a dataset is proven by Young and Householder [206], who also provide a formula for the dimensionality of that space. We may note that this data matrix is not unique. Indeed, all transformations of a dataset by an isometry produce the same distance matrix. In a Euclidean space, isometries are all combinations of rotations, symmetries and translations.

1.1.2 Neighbourhood Ranks

Neighbourhood ranks reduce the information of distances for a given dataset to an ordering, considering independently each row of the distance matrix. The rank ρ_{ij} describes the position of point ξ_j in the neighbourhood of point ξ_i, that is its place in the sorting of data points by their distance to point ξ_i. Replacing distances values by their ranks ensures more robustness to the phenomenon of norm concentration detailed in Sect. 2.1. Formally, a rank ρ_{ij} indicates that point ξ_j is the ρ_{ij}th nearest neighbour of point ξ_i. By convention, we set $\rho_{ii} \triangleq 0$.

For each data point ξ_i, we define the neighbourhood permutation $\tilde{v}_i : [\![0, N-1]\!] \longrightarrow [\![1; N]\!]$ as the mapping returning for a given rank κ, the index j of the κth nearest neighbour of ξ_i in that space. Namely, $\tilde{v}_i(\kappa)$ is the index so that $\xi_{\tilde{v}_i(\kappa)}$ is the κth nearest neighbour of ξ_i. We may note that $\tilde{v}_i(\rho_{ij}) = j$ and that, using the bijectivity of the permutation, $\rho_{ij} = \tilde{v}_i^{-1}(j)$ (which may be an alternative definition of ranks).

We also define κ-neighbourhoods $v_i(\kappa)$ as the set of indices of the κ nearest neighbours of i. This may be formally defined based on ranks as $v_i(\kappa) = \{j \neq i \mid \rho_{ij} \leqslant \kappa\}$, or as the image by the neighbourhood permutation $\tilde{v}_i$ of the set $[\![1; \kappa]\!]$,

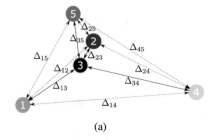

$j = \tilde{v}_3(\rho_{3j})$	Δ_{3j}	$\rho_{3j} = \tilde{v}_3^{-1}(j)$
1	74.2	3
2	29.3	1
3	0	0
4	126.4	4
5	54.4	2

(a)   (b)

Fig. 1.1 Illustration of the relation between pairwise distances Δ_{ij}, neighbourhood ranks ρ_{ij} and neighbourhood permutations $\tilde{v}_i$ on a toy example of abstract metric data, focusing on point $i = 3$. (**a**) Points and their pairwise distances Δ_{ij}. (**b**) Values for $i = 3$

namely $v_i(\kappa) = \tilde{v}_i(\llbracket 1; \kappa \rrbracket)$. The link between distances, neighbourhood ranks and neighbourhood permutations is illustrated Fig. 1.1, for an abstract metric dataset.

1.1.3 Embedding Space Notations

In the context of dimensionality reduction, all these notions are symmetrically defined in the data space $\mathcal{D}$ and in an embedding space $\mathcal{E}$, usually equipped with the Euclidean distance. Hence, embedding space distances D_{ij}, ranks r_{ij}, neighbourhoods $n_i(\kappa)$ and neighbourhood permutations $\tilde{n}_i$ are the counterparts for the embedded points x_i of Δ_{ij}, ρ_{ij}, $v_i(\kappa)$ and $\tilde{v}_i$ defined for the data points ξ_i. Note that the ith point is often referred to as point i, that point being characterized both by its coordinates ξ_i in the data space and x_i in the embedding space. The neighbourhood relation between a point i and a point j is denoted $i \sim j$.

1.1.4 Multidimensional Data

Multidimensional data (also called feature data or tabular data) correspond to a set of N data points (or feature vectors) ξ_i in a high dimensional vector space $\mathcal{D}$. This data space (or feature space) $\mathcal{D}$ of dimensionality δ often corresponds to $\mathbb{R}^\delta$. A multidimensional dataset may be stored in a *data matrix* Ξ of size $N \times \delta$. The element (i, k) of that matrix, denoted ξ_{ik}, contains the value of the kth variable for the ith data point ξ_i.

Multidimensional data is thus the natural format for treating data tables, which are the basic element of relational databases (e.g., SQL databases). Indeed, those tables are organized by rows and columns, each row corresponding to an instance, and each column being associated with an attribute (or feature) of that instance. In statistics, those instances are also called individuals or observations.

The features are either quantitative, such as numerical or ordinal variables, or qualitative, as for categorical or boolean variables. Yet, all these types of variables may be stored in a common numerical matrix with, for example, ordinal variables represented by successive integers, boolean variables by 0 and 1 values and categorical variables represented by several boolean variables (one by category), each indicating whether the observation belongs to that category [179].

For a data matrix Ξ, an associated distance matrix Δ_Ξ may be obtained by choosing a specific metric Δ on the data space. Dimensionality reduction seeks to convert metric data into multidimensional data in a low dimensional space, thus leading to a set of N embedded points x_i in a low dimensional embedding space $\mathcal{E}$ of dimensionality d.

Minkowski Distances
Multidimensional data points may be compared coordinate by coordinate, for instance using the family of Minkowski distances Δ_p. These distances are based on the associated Minkowski norms, also called L_p norms, parametrized by $p \in [1; +\infty[$. For two vectors ξ_i and ξ_j of $\mathbb{R}^\delta$, it is computed based on their coordinates ξ_{ik} and ξ_{jk}, rendering it basis dependent. Its general expression is:

$$\Delta_p(\xi_i, \xi_j) = ||\xi_i - \xi_j||_p = \left(\sum_{k=1}^{\delta} |\xi_{ik} - \xi_{jk}|^p \right)^{1/p}.$$

For $p = 2$ it corresponds to the standard Euclidean distance:

$$\Delta_2(\xi_i, \xi_j) = \sqrt{\sum_k (\xi_{ik} - \xi_{jk})^2}.$$

Its associated norm is the only one of the family that may be linked with an inner product and that is invariant to multiplication by an orthogonal matrix (i.e. change of basis by rotations and/or other symmetries). Hence, the unit ball for that norm (see Fig. 1.2) is rotation invariant.

For $p = 1$ it corresponds to the "Manhattan" distance, which sums the absolute values of differences:

$$\Delta_1(\xi_i, \xi_j) = \sum_k |\xi_{ik} - \xi_{jk}|. \tag{1.2}$$

For distances between regularly sampled curves (small and uniform step), it is approximately proportional to the area between the curves.

When p increases, the Minkowski distance progressively gives higher importance to coordinates for which there is the most difference between the vectors. It asymptotically tends to the Chebyshev distance as $p \to +\infty$, whose associated norm finds the maximum of coordinates absolute values:

Fig. 1.2 Unit balls for the family of Minkowski norms (and pseudo-norms). We may note that for a pseudo-norm, the triangle inequality property is equivalent to the convexity of its unit ball (as shown in Annex A.1).

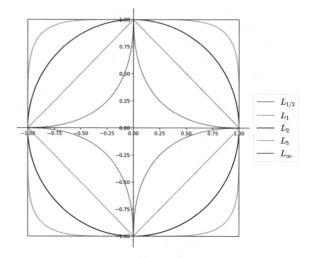

$$\Delta_\infty(\xi_i, \xi_j) = \max_k |\xi_{ik} - \xi_{jk}|.$$

"Fractional" values of p (lower than 1) have been used [1] to improve the contrast of the measured dissimilarities. This allows to alleviate the problem of norm concentration detailed in Sect. 2.1. Yet, $p < 1$ induces a loss of the triangle inequality property. A pseudo-norm is also defined for $p = 0$, counting the number of non-zero coordinates of the vector. It is not actually a norm, since it does not satisfy homogeneity (as it is invariant to multiplication by a non-zero scalar). However, it defines the Hamming distance, used to compare vectors of boolean or categorical variables, which counts the number of non-equal coordinates between the two vectors:

$$\Delta_0(\xi_i, \xi_j) = \left| \{k \in [\![1; \delta]\!] \mid \xi_i \neq \xi_j\} \right|,$$

with $| \cdot |$ the cardinal of a set.

1.1.5   Sequence Data

A sequence θ_i is a collection of values $\theta_{ik} \in \mathcal{V}$ indexed by an auxiliary variable $t_i \in \mathcal{T}$, that is so that $\theta_{ik} = \theta_i(t_{ik})$ for $k \in [\![1; \delta_i]\!]$. $\mathcal{V}$ is the value space, and $\mathcal{T}$ is the index space. The auxiliary variable may induce strong correlations between successive values of a given sequence. Sequences include time series (indexed by time), spectra (indexed by frequency) or angular data (indexed by angles). They also extend to sequences of symbols (i.e. categorical variables), such as character strings or DNA sequences. The indexing variable t_i may also be bi-dimensional or three-dimensional as for images (indexed by pixel position) or spatial data (indexed by spatial coordinates), though it loses the order relation. Conversely, the values of the sequence may be multidimensional as for an I–V curve, which is a time series (see Sect. 8.2). Sequences with common indexing (e.g., time series, character strings or images of common size) may be treated as feature vectors. This corresponds to mapping for each index value t_{ik} (e.g., the time stamp, character position or pixel) the sequence value $\theta_i(t_{ik})$ to an associated feature ξ_{ik}.

Sequence Alignment: Edit Distance and Dynamic Time Warping
Minkowski distances, applied to multidimensional data, compare data points coordinate by coordinate. Yet, with sequence data, this is only relevant if values may be aligned along a common index. Sequence alignment methods allow to compare sequences with different indices, by matching together similar sub-sequences.

Considering character strings of same length, the Hamming distance (Eq. (1.2)) between two strings accounts for the minimum number of character substitutions required to transform one string into the other. It is extended to strings of arbitrary length by the Levenshtein distance which considers the smallest number of substitutions, insertions and deletions. Insertion and deletion allow to align sub-strings that are very close in terms of the Hamming distance. The broader family of edit distances allow to attribute a different weight to insertion, deletion and substitution operations (which may lead to losing the symmetry property of a distance). Such distances are used for instance in phylogeny for comparing proteins (sequences of amino-acids) [117]. In that case, the weight associated to any given operation accounts for its effective molecular cost, as for the Point Accepted Mutation (PAM) distance [46, 66].

For time series, a similar principle is used by Dynamic Time Warping (DTW) [156] which searches for the best time warping finding the minimum deviation between two series. Figure 1.3 illustrates the alignment of points by DTW as opposed to Euclidean distances. For the Euclidean distance, values of common time index are compared together (Fig. 1.3a). The sequences not being of the same length, some values of the longest sequence must be removed to allow comparison. Conversely, for the DTW, points are aligned so as to compare the global evolution of each curve. The triangle inequality property does not hold for the DTW distance, but empirical results obtained for sets of word recognition sequences show that it is loosely satisfied (i.e. satisfied in a statistically significant number of cases) [190].

1.1.6 Network Data

Networks data characterize relations between instances, as can be stored in a relation table in relational databases. As such, they can be modelled by a graph (as formally

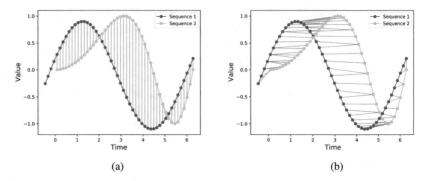

Fig. 1.3 Alignment of two time series for Euclidean distance and Dynamic Time Warping (DTW) computation. For the Euclidean distance, the values of the two series are compared for common time steps. Hence, some values (in red on sequence 1) must be removed since the time scales are not common for this part. For the DTW, time is warped so that values are compared between different time steps, allowing to identify common trends in the evolution of the sequence value. (**a**) Euclidean distance alignment. (**b**) DTW alignment

defined by Definition 1.4). They may either be hierarchical data (tree structures) or relational data (graph structures).

Definition 1.4 A weighted directed graph (or digraph) $G = (V, E, W)$ is composed of:

- V the set of N vertices,
- $E \subseteq V \times V$ the set of directed edges with cardinal $|E| \leqslant N^2$,
- W the set of weights associated to the edges.

The vertices $i \in V$ of that graph correspond to instances and edges $(i, j) \in E$ to the relations existing between the instances i and j. The associated weights w_{ij} characterize those relations. They may, for example, be measures of similarity γ_{ij} or measures of dissimilarity Δ_{ij}. A graph weighted by similarities may be represented by its adjacency matrix whose element (i, j) contains the weight w_{ij} if the edge (i, j) exists and 0 otherwise. For non-complete graph, that matrix is sparse. This representation could be adapted to graphs weighted by dissimilarities by denoting non-existing edges with elements equal to $+\infty$.

Graph Distances
Weights of a graph often define similarities or dissimilarities between some pairs of vertices. Graph distances rely on this sparse information to define a full distance matrix Δ measuring dissimilarity between all pairs of vertices.

Shortest path distances [175] find the path of minimum length between two vertices in the graph weighted by dissimilarities. Conversely, in graphs weighted by similarities, distances tend to rely on random walks. Those random walks take a random path resulting from successive random transitions, where the probability of transitioning from a vertex i to any other vertex j is proportional to the weight w_{ij}.

This is used both by the commute time distance [149] and the diffusion distance [34, 130]. The commute time distance Δ_{ij}^{CT} measures the average number of steps required to go from i to j and back in the random walk. Conversely, the diffusion distance $\Delta_{ij}^{diffus}(t)$ considers the probability of arriving at any given point k in t iterations either starting from point i or from point j. The distance is then computed by comparing the distributions of those probabilities over all points k. Diffusion distances highly depend on the choice of the parameter t [130].

1.1.7 A Few Multidimensional Datasets

This section provides some examples of multi-dimensional datasets. Those are either geometric datasets, defined by their structure in the data space, or real datasets obtained by measures of some physical quantities.

1.1.7.1 Geometric Datasets

Geometric datasets are often designed with a desired data structure, to illustrate some characteristics of a method of data analysis (as listed in Sect. 1.2). Indeed, in the case of multidimensional data, one may construct a set of points in an abstract feature space, satisfying certain geometrical or topological properties. This can be done regardless of the meaning of data space axes, which can be thought of as spatial coordinates. If this space is of very low dimensionality (i.e. $\leqslant 3$), the data structure may appear in a visual and intuitive way.

We can, for example, list the following datasets, commonly used to illustrate dimensionality reduction methods performances:

- **Swiss roll** (Fig. 1.4a) [175]: a spiral extruded along the 3rd orthogonal direction, thus constituting a two-dimensional surface in a three-dimensional space. Local preservation of the similarities requires to unroll that surface. It may be mapped to the plane in an intuitive way by unrolling the surface.
- **Two rings** (Fig. 1.4b) [54]: two one-dimensional entangled circles in a three-dimensional space. A simple representation of the local structure consists in two separate circles, but it does not render the global topological structure.
- **Two open boxes** (Fig. 1.4c) [114]: the surface of two cubes each with one missing facet (different for each cube). Each cube can be represented in two dimensions by unfolding it. Yet their respective orientation must be changed, since the missing facet is oriented differently for each cube.
- **Sphere** (Fig. 1.4d) [189]: the simple two-dimensional surface of a sphere in a three-dimensional space. That dataset may not be perfectly represented in a two-dimensional plane without distortion, even considering the only the local structure. Hence, it allows to illustrate the type of distortion favoured by a DR

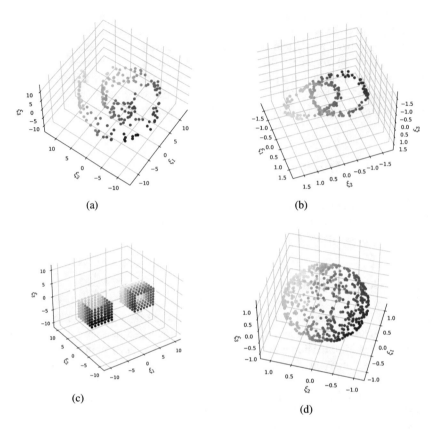

Fig. 1.4 Example of three-dimensional toy datasets, coloured by mapping each axis to a coordinate of the RGB colour space [185]. (**a**) Swiss roll. (**b**) Two rings. (**c**) Two open boxes. (**d**) Sphere

technique. The sphere dataset may be generated by normalizing the samples drawn from an isotropic three-dimensional Gaussian distribution [55].

- Smiley dataset [173]: groups of points shaped as parts of a face (two Gaussian-distributed eyes, a trapezoid nose and a parabolic mouth). This is used to represent the preservation of local and global properties of the structure. Indeed, if only the local properties are considered, the face parts are preserved but the relations between them are distorted (leading to a Picasso-like representation of the face).

1.1.7.2 Real Datasets

Real datasets illustrate effective applications of data science. They tend to be of higher dimensionality and to systematically contain some noise. The data points are also sorted into categories, called classes, based on semantic information. Some examples are:

- Iris dataset [67]: a historical multidimensional dataset containing 150 instances of iris flowers of three varieties (setosa, versicolor and virginica) and characterised by four anatomic measurements (petal and sepal widths and lengths). The classes (varieties) are balanced with 50 instances each.
- Wine dataset [58]: 178 samples of Italian wines characterized by their quantities of 13 chemical constituents. They are split between three classes depending on the cultivar used for growing the wine.
- Oil flow [18]: 500 measurements of 12 variables characterizing the flow of oil in a pipeline, obtained through numerical simulation, and separated into three classes corresponding to the possible configurations of the three phases of the flow (oil, water and gas). Those configurations are "homogeneous", "annular" and "stratified". The input parameters of numerical simulations determine the respective fractions of oil, gas and water in the pipeline. Hence, there are two degrees of freedom in those parameters.
- Columbia Object Image Library (COIL-20) [136]: 1 440 gray-scale pictures of size 128×128 pixels (16 384 dimensional feature vectors), representing 20 different objects rotated around a vertical axis each photographed with 72 angular steps with 5° intervals. This leads to 20 balanced classes each containing 72 elements. This is a very high dimensional dataset, whose classes have only one degree of freedom (the angle of the rotation). Figure 1.5 presents some of those images.
- Handwritten digits [4, 58]: 3 823 instances of 8×8 (64-dimensional feature vectors) gray-scale images of the 10 digits written by hand. The classes are relatively balanced with from 376 to 389 instances by class. Some digits images are shown Fig. 1.6. The MNIST (Mixed National Institute of Standards and Technology) [103] is very similar to the digits data set, but with different angles of the digits (additional rotation).
- Isolet 5 [58, 63]: 30 speakers each uttering twice all of the 26 letters of the Latin alphabet with English pronunciation, producing 26 balanced classes of 60 instances each (except for one missing M). The 1 559 instances are described by 617 features. These pronunciations may be transcribed in the International Phonetic Alphabet (IPA) as A [eɪ], B [biː], C [siː], D [diː], E [iː], F [ɛf], G [dʒiː], H [eɪtʃ], I [aɪ], J [dʒeɪ], K [keɪ], L [ɛl], M [ɛm], N [ɛn], O [oʊ], P [piː], Q [kjuː], R [ɑr], S [ɛs], T [tiː], U [juː], V [viː], W [ˈdʌbəljuː], X [ɛks], Y [waɪ], Z [ziː]).

Fig. 1.5 Representative images of the COIL-20 dataset

Fig. 1.6 Representative images of the handwritten digits dataset

For images datasets, such as handwritten digits and COIL-20, presenting the images may give an intuitive sense of the Euclidean distance between two instances in the data space.

1.2   Automated Tasks

Data science provides algorithmic solutions to perform data analysis in an auto-mated manner. Those tasks generally consist in estimating the unknown value of target variables (or explained variables) based on known features (or explanatory variables). This amounts to associating to a feature vector ξ (usually numerical) a label L (categorical target variable) or a real-valued vector x (numerical target variables). Such associations defines a mapping

$$\widehat{\Omega} : \begin{cases} \mathcal{D} \longrightarrow \mathcal{L} \\ \xi \longmapsto L \end{cases} \quad \text{or} \quad \widehat{\Phi} : \begin{cases} \mathcal{D} \longrightarrow \mathcal{E} \\ \xi \longmapsto x \end{cases}$$

where $\mathcal{D}$ is the data space, $\mathcal{L}$ a label space and $\mathcal{E}$ a real vector space. Note that hard labels L (categorical variables) may be replaced by soft labels $\tilde{L}$ each corresponding to the distribution of probabilities that the point associated to the soft label belong to a category or another. This distinction of hard and soft labels is detailed more in Sect. 6.1. We distinguish two main usages of these algorithmic tools:

- *Descriptive analysis* (or exploratory analysis) consists in identifying structures in the data, such as categories or latent variables explaining the observed variability of data. Its core tasks are clustering and dimensionality reduction. Those do not require any prior knowledge about the target variable, thus constituting unsupervised tasks. As such, they may lead to unexpected discoveries about the data.
- *Predictive analysis* aim at estimating or predicting the value of a well-identified target variable based on the available features. It relies on the definition of a model, fitted using training data for which both the features and target variables have known values. Hence, its core tasks, that are classification and regression, are supervised.

Table 1.1 provides a taxonomy of tasks based on the double distinction by supervision of the task and type of target variable.

Table 1.1 Taxonomy of data science tasks (inspired from [171])

| | | Task supervision | |
		Unsupervised	Supervised
Target variable type	Categorical	Clustering (Sect. 1.2.2.1)	Classification (Sect. 1.2.4.1)
	Numerical	Dimensionality reduction (Section1.2.3.1)	Regression (Sect. 1.2.5)

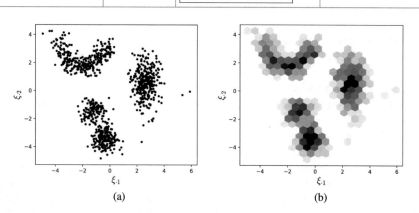

(a)   (b)

Fig. 1.7 Simple two-dimensional data structure generated by a mixture of a uniformly sampled parabola with added Gaussian noise (top left), two isotropic Gaussian components (bottom left) and one anisotropic Gaussian component (right), and a few points uniformly sampled in the square. Left panel shows points sampled from the distribution, while right presents an empirical estimation of the underlying distribution. (**a**) Data set. (**b**) Underlying distribution

1.2.1   Underlying Distribution

A dataset is constituted of points $\{\xi_i\}$ organized according to underlying data structures. These points are often assumed to be independent and identically distributed (i.i.d hypothesis). This means that they are samples independently drawn from the same probability distribution defined over the data space $\mathcal{D}$ [81]. In practice, however, that unknown theoretical continuous distribution can only be estimated through the available discrete samples contained in the available dataset. We present in Sects. 1.2.2.1 and 1.2.3.1 several kinds of data structures with theoretical definitions (mainly based on this underlying distribution) and practical data science methods used to identify those structures. Figure 1.7 provides an illustration based on a two-dimensional toy dataset, with the data points (Fig. 1.7a) and an empirical estimation of the underlying distribution (Fig. 1.7b).

1.2.2   Category Identification

This Section concerns the detection of clusters, hierarchies and outliers in the data, thus regrouping most of the data points into categories sometimes organized into a hierarchical structure.

1.2.2.1   Clusters and Flat Clustering

Clusters are groups of points that are similar to each other and dissimilar to points from other clusters. In terms of the underlying distribution, a cluster constitutes a connected area of high density around a mode of the distribution. Clusters may be determined automatically by clustering algorithms providing a flat clustering, or visually relying on the ability of human cognition to identify groups (see Gestalt laws of proximity and continuity detailed Sect. 1.3.2). Indeed, by looking at Fig. 1.7a, the reader gets an intuitive idea of what the clusters are for this dataset (a priori close to the automatic clustering of Fig. 1.8b).

Clustering algorithms identify a latent categorical variable indicating the cluster to which a given point belongs. Namely, they determine a mapping $\Omega : \mathcal{D} \longrightarrow \mathcal{L}$ assigning each data point ξ_i to a category with a label $L_i = \Omega(\xi_i)$. The number of clusters, that is the number of possible values of that categorical variable, is a key parameter for a flat clustering. We may distinguish two main approaches for clustering of multidimensional data: the parametric approach used by partitioning algorithms and the density-based approach. For network data, the equivalent of clustering is community detection. In terms of graphs, communities (i.e. clusters) may be defined as groups of vertices linked together by many edges and linked to their surroundings by less edges [19].

Parametric Clustering
Partitioning algorithms, such as k-means [118] and k-medoids [96] split the space into k convex regions parametrized by associated prototypes. Indeed, they assign

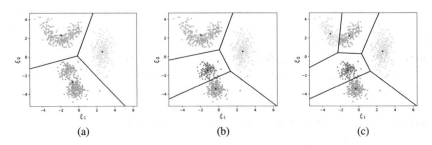

| (a) | (b) | (c) |

Fig. 1.8 k-means clustering for the two-dimensional toy dataset presented Fig. 1.7, for several values of k. The black points and lines represent the clusters prototypes and the boundaries of their Voronoï cells. (**a**) $k = 3$. (**b**) $k = 4$. (**c**) $k = 5$

each point of the datase to one of the clusters, so as to minimize the distances separating points from their clusters prototype. This prototype, which is respectively a centroid for k-means and a medoid for the k-medoids, provides a central tendency of the cluster. Formally, those algorithms seek the clustering that minimizes the cumulated Fréchet variance of all clusters, measured around their respective Fréchet means, which is the aforementioned prototype.

The most common approach to solve this problem, that are Lloyd algorithm for k-means, andt Partitioning Around Medoids (PAM) for k-medoids, are organized in the following steps. First, the positions of the prototypes are initialized (randomly or using seeds selection). Then, at each step, the points are assigned to the cluster of their closest prototype and the prototypes are recomputed based on the new points of the clusters. The process iterates until convergence of the prototypes positions. Hence, this partitioning may be seen as splitting the space into the Voronoï cells associated with the prototypes (as shown Fig. 1.8). As such, partitioning methods assign every points to a cluster and do not detect outliers.

Gaussian Mixture Model [126] clustering provides an explicit parametric model of the distribution as a mixture of Gaussian components. These components are fully defined by their means and covariance matrices (allowing anisotropic clusters), each component corresponding to one cluster. The optimization of those parameters is performed using the Expectation Maximization (EM) algorithm, in order to maximize the likelihood of the model. The final clustering may either attribute each point to its more likely component (hard clustering), or give a probability distribution indicating the probability that a point belongs to one cluster or another (soft clustering).

Those parametric clustering methods assume a given shape of the clusters, and rely on an input k that is the expected number of clusters. Figure 1.8 illustrates the result of k-means algorithm for the toy dataset for several consecutive values of k.

Density-Based Clustering

The density-based approach is non-parametric and does not assume a specific shape of clusters. It relies on the estimation of the underlying distribution. As such, it tends to require more points than parametric methods, in order to get a good assessment of that distribution. The number of clusters is an output of those algorithm. It depends on specific parameters used to identify the regions around theoretical modes of the distributions.

Density-Based Spatial Clustering of Applications with Noise (**DBSCAN**) [61] looks for clusters that are connected components of a super-level set of the probability density function of the underlying distribution. These are connected regions of the data space for which the density of points is above a certain density threshold. This may be apprehended based on Fig. 1.7b, by looking at areas darker than a certain level of shade. In practice, **DBSCAN** finds these regions by identifying core points, which are the points that have at least a fixed number of neighbours within their ω-neighbourhood, with ω being a fixed radius. Clusters are thus connected components of the ω-neighbour graph of core points. For non-core

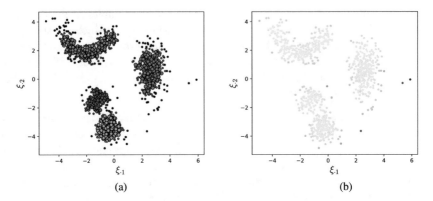

Fig. 1.9 Density-based clustering and outlier detection for the toy dataset from Fig. 1.7a. DBSCAN clustering (**a**) identifies four clusters distinguished by colours. Core and non-core points of these clusters are distinguished by marker size, while outliers are coloured in black. The LOF measure (**b**) is displayed using a YlOrRd colourmap [27] identifying the three clear outliers. (**a**) DBSCAN. (**b**) LOF

points, a distinction is made between the border points and outliers. Border points are in the neighbourhood of a core point and may be assigned to its cluster. Inversely, outliers are in none of those neighbourhoods, and thus in regions considered as of low density. Figure 1.9a presents clustering of the toy dataset by DBSCAN with a specific set of parameters leading to the detection of four clusters.

The mean-shift algorithm [38] is closely related to the density-based definition, since it estimates the density function using the non-parametric Kernel Density Estimation (KDE). Yet, it does not explicitly defines a density threshold. To determine the modes, it iteratively displaces points in space in the direction of the gradient of the estimated density function. At each step, the density is re-estimated with the new positions of the points, so that points converge towards their closest mode. Thereby, the number of obtained modes depends on the bandwidth parameter and of the kernel chosen for the KDE. A smoother density function (large bandwidth) leads to less modes, and thus less clusters.

1.2.2.2   Outliers and Outlier Detection

Outliers are points far from the others or found in regions of space with very low density regarding the probability distribution. In the process of clustering the points based on density, outliers correspond to the points belonging to none of the clusters, as in DBSCAN example presented Fig. 1.9a. They may also be identified automatically using specifically designed methods for outlier detection, such as the Local Outlier Factor (LOF) [26]. A high value of that factor indicates that the density is lower for a point than for its neighbours meaning that it is lying out of the clusters. Figure 1.9b illustrates the distribution of the LOF for the points of the toy dataset.

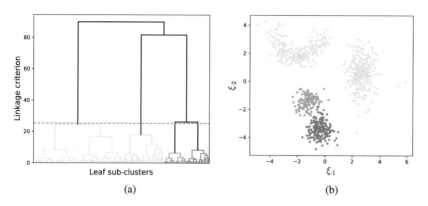

Fig. 1.10 Hierarchical structure obtained by an agglomerative clustering with Ward criterion. The dendrogram on the left insert represents the hierarchical structure obtained by successive merging of sub-clusters. The ordinate of that figure indicates the value of the linkage criterion associated to each merge. By selecting a certain level of that hierarchy (horizontal dashed line) a flat clustering is obtained, as shown on the right insert. (**a**) Dendrogram. (**b**) Hierarchical clustering

1.2.2.3   Hierarchies and Hierarchical Clustering

The grouping of points into clusters in a flat clustering is defined at a given scale, either depending on a number of sought components (partitioning) or on a density threshold (density-based). Yet, clusters at a given scale may be split into several sub-clusters (increasing the number of components or density threshold) or merged with other similar clusters forming a new higher scale grouping (decreasing the number of components or the density threshold). This multi-scale organization of groups constitutes a hierarchy. This structure may be automatically identified using hierarchical clustering algorithms, illustrated by Fig. 1.10 for the toy dataset. It is often presented as a tree called dendrogram (see Fig. 1.10a).

Hierarchical clustering [196] may either be ascending or descending. Ascending (or agglomerative) methods iteratively merge the clusters, starting from an initial state where each point constitutes its own cluster. Conversely, descending (or divisive) methods successively split the clusters, starting from a cluster containing the whole dataset. At each iteration, the best merge or split is selected based on a specific criterion (e.g., single linkage, mean linkage, complete linkage, Ward [134]).

Hierarchical clustering may be used to provide a flat clustering, by selecting a certain level of the tree, considering that some scales are more natural than others for the specific structure of a given dataset. Figure 1.10b presents the flat clustering obtained by selecting a specfic level of the agglomerative clustering presented Fig. 1.10a. We may note that this clustering corresponds to a value of the linkage criterion.

HDBSCAN (for Hierarchical Density-Based Spatial Clustering of Applications with Noise) [30] adapts this hierarchical approach to density-based clustering. It also proposes a way to obtain a flat density-based clustering by selecting different levels of the tree structure, providing clusters of varying densities (as opposed to DBSCAN).

1.2.3   Data Manifold Analysis

This Section presents the techniques for identifying the structure of a data manifold, either by extracting latent variables locally parametrizing this manifold, or by identifying its topological structure.

1.2.3.1   Latent Variables Extraction and Manifold Learning

In the *i.i.d* hypothesis, the support of the theoretical probability distribution generating data points $\{\xi_i\}$ is considered as a manifold M immersed in the ambient data space $\mathcal{D}$ [9, 81]. The repartition of points along a manifold may be explained by the strong dependency between data space variables. In addition, one may assume that all these variables are local functions of a few independent latent variables with an additional noise [176], thus constituting a low-dimensional structure. That noise may induce small variations around the smooth structure of that manifold. Note that the manifold hypothesis may extend to datasets that are not generated by random processes. For instance, for the two open boxes and COIL-20 datasets (see Sect. 1.1.7), data lie on a low-dimensional manifold which is regularly sampled, and not randomly sampled.

Dimensionality Reduction (DR) in general aim at finding a mapping $\Phi : \mathcal{D} \longrightarrow \mathcal{E}$, that associates each data point ξ_i to a point $x_i = \Phi(\xi_i)$ in a low dimensional embedding space $\mathcal{E}$. A key parameter of dimensionality reduction is the embedding dimensionality d (i.e. the dimensionality of $\mathcal{E}$). We distinguish here two sub-cases of DR: manifold learning and spatialization. The ideal goal of *manifold learning* is to extract latent variables parametrizing the manifold, which explain the variability of data. Those hypothetical variables may also be referred to as curvilinear components of the manifold [54]. In that case, the embedding dimensionality defines the number of variables to extract. A possible value for that parameter is the intrinsic dimensionality, which corresponds locally to the number of curvilinear components required to parametrize the manifold (see Sect. 2.2). Manifold learning may be used as a pre-processing step for other machine learning applications (e.g., classification or clustering), in order to mitigate the curse of dimensionality [155], to compress the data [179], or to filter out the noise [176]. Inversely, *spatialization* aims at providing a visual representation of high-dimensional data (see Sect. 1.3.2). As a result, the embedding dimensionality is constrained by the perceptual capabilities of the data analyst, limiting the number of dimensions to at most three for visualization with only one scatter plot. Satisfying this strong constraint on dimensionality often requires distortions of the underlying data structure. Note that the equivalent of DR for network data is graph embedding (also called graph layout).

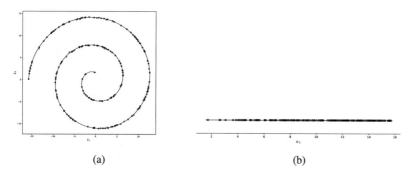

Fig. 1.11 Intrinsically one-dimensional spiral manifold in a two-dimensional space and its embedding by an ideal manifold learning technique (extracted curvilinear abscissa). (**a**) Spiral manifold. (**b**) Ideal embedding of the spiral manifold

Figure 1.11a illustrates the case of a spiral shaped manifold in a two-dimensional (2D) data space. Figure 1.11b presents the values of the curvilinear abscissa along that spiral that should be extracted by a manifold learning method. This constitutes a one-dimensional (1D) embedding of the data set.

Geodesic Distances
Assuming that data live in a manifoldt immersed in the ambient data space, it seems more relevant to measure distances along that manifold rather than in the ambient space. Formally, these geodesic distances (in the corresponding Riemannian metric) measure for each pair of point, the length of the shortest path between them on the manifold. Yet, this theoretical underlying manifold is only known through the observed data points, so that the geodesic distances must be approximated based on ambient space distances.

Nonlinear manifolds are locally curved, which can lead to a folded global structure in the ambient data space, as for the one-dimensional spiral manifold presented Fig. 1.12. Hence, points that are apart in the manifold may be far closer in the ambient space. At sufficiently small scales, the manifold is almost linear, which reduces the effect of curvature. Thereby, the short ambient space distances are considered to be the most reliable to model distances along the manifold. This lead to computing the estimated geodesic distances with the following steps [175]:

- build a proximity graph (such as an undirected κ nearest neighbours graph) weighted by the ambient space distances between the vertices,
- compute the geodesic distances between all pairs of points as the shortest path distances in that graph.

The parameter κ defines the scale at which the manifold is assumed locally linear. Its choice is subject to the following trade-off: it must be sufficiently high to avoid disconnected components in the κ-nearest neighbours graph, and sufficiently small to prevent manifold short-circuiting [130, 184]. The latter phenomenon, illustrated Fig. 1.12, corresponds to spurious paths connecting separate regions of the manifold.

Fig. 1.12 Illustration of the
short-circuiting problem on
the one-dimensional spiral
manifold with Gaussian noise

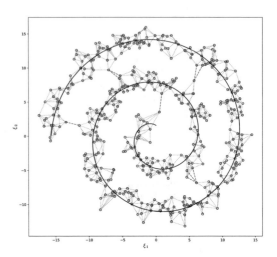

It leads to strong underestimation of the geodesic distance between those regions.
The risk of manifold short-circuiting increases with the global folding of the
manifold and the presence of noise.

Other measures of distances along the manifold may be obtained by constructing
a proximity graph weighted by similarities and computing the commute time
distance or diffusion distance detailed in Sect. 1.1.6 [34, 130, 149]. The use of
those random walk based graph distances reduces short-circuiting compared with
the shortest path distances. Indeed, the spurious paths have a low probability in the
collection of all possible paths and their effect will tend to vanish in a random walk
in comparison to other longer paths along the manifold [130].

1.2.3.2 Continua and Topology Learning

Another way of studying the global structure of data is topology learning, which
may help to identify some topological properties of the manifold (e.g., the number
connected components, holes and their dimensions, etc.). For instance, in the toy
dataset presented Fig. 1.7, we may identify a manifold constituted of three or
four connected components (depending on the density threshold for defining the
support). Topology learning is the goal of the Generative Gaussian Graph (GGG)
[9, 73], which models the manifold by a set of points and segments with added noise.
This is proposed as a tractable version of a generative simplicial complex modelling.
These points and segments may be initialized as a set of prototypes of the dataset and
the edges of the Delaunay graph between those prototypes. They are then optimized
by Expectation Maximization (EM) of the Gaussian mixture generated by all points
and segments. In this approach, the segments form one-dimensional (1D) structures
at the interface between clusters, that we call here continua.

1.2.4 Model Learning

1.2.4.1 Classification

Classification consists in assigning to any data ξ a category, called class, with a label L. These classes are often human-defined concepts. To infer categories, classification relies on data instances $\{\xi_i\}$ of a training set for which the true class is provided through annotations $\{L_i\}$. As such, classification assumes that there exists a relation linking the classes to the data features. A classifier then seeks to define decision boundaries partitioning the space into regions, each corresponding to a separate class. Classification differs from clustering which guesses the categories based only on the data structure, hence giving no specific meaning to those categories. Some of the many classification techniques existing in the literature are described below.

k-Nearest Neighbourst (k-NN) classifiers [43] estimate the class of a test point based on the class of its nearest neighbours. Its performances mainly depend on the choice of an appropriate number of neighbours k as well as a relevant metric. A too small value of the hyper parameter k may lead to over-fitting, while a too high value may induce under-fitting.

Support Vector Machines (SVM) [42] search for the hyperplane best separating the classes. It is best-suited for linearly separable classes. Yet, kernel approaches [28] allow to benefit from the curse of dimensionality, providing linear separability in a kernel space, which amount to non-linear decision boundaries in the original data space.

Random forests [25] use ensemble learning, combining the classification obtained by several decision trees. These individual trees (base learners) are built using a certain randomly sampled set of the training points and of the data features, thus allowing more generalizability than a unique decision tree built using a specific dataset.

Artificial neural networks have high generalization capabilities. They use a combination of multiple neurons (or units), often organized in several layers. Each neuron combines a linear form of its entries and a scalar non-linear activation function. The simple fully connected architecture may be replaced by more specific architectures such as convolutional neural networks (for images and sequences) [102] or recurrent neural networks.

1.2.5 Regression

Regression methods allow to define a mathematical model $\widehat{\Phi}$ predicting real-valued target variables x based on data features ξ. This model is obtained by fitting it to the training set for which both the features $\{\xi_i\}$ and target $\{x_i\}$ are known. Common approaches to regression are kernel regression, that performs weighted average of the value for the near neighbours [135, 197], linear and polynomial regression, which fit a polynomial model, random forests of regression trees [25] and artificial neural networks.

1.3   Visual Exploration and Visual Encoding

Dimensionality reduction provides low-dimensional representation of high dimensional data. With a very low dimensionality, this representation, called map, may be employed for visual exploration of the dataset. This visual component is essential to incorporate human cognition in the strongly automated manipulation of data. As such, dimensionality reduction, whose output is usually represented with scatter plots, is not the only solution, and several other approaches may be used to encode data visually.

1.3.1   Human in the Loop Using Graphic Variables

Visual data analytics allows to integrate human cognition into a data science pipeline, through graphic representation either of data. These exploratory steps may help the analyst to *forge an intuition* about the data structure, in order to assist the choice of machine learning algorithms to perform a predictive analysis, or to help *form hypotheses* that can then be tested by confirmatory analysis (i.e. statistical tests).

In this human in the loop framework, cognitive load, namely the amount of an analyst's working memory required by a task, is predominant. In this respect, the cognitive load in visual exploration steps is dual with the computational load of machine learning algorithms. Indeed, interpreting raw data presented in the form of huge numeric tables may be considered as an intractable task for an analyst, due to the enormous cognitive load involved. In contrast, visually encoding information by mapping variables (i.e. the columns of the aforementioned table) to perceptual channels significantly reduces this cognitive load, allowing to obtain information at a glance. This difference may be explained by the sequential treatment of the textual transcriptions which can only convey unidimensional (1D) information; as opposed to visual perception which provides a three dimensional (3D) information treated in parallel [17].

Visual representation of data requires to define a visual encoding [163], which means associating each variable to a graphic variable. The graphic variables may be interpreted by the human brain with varying levels of accuracy [17]. Some are only suited for representing qualitative variables, while others may also represent quantitative variables. The following categories are identified in [23]:

- Position, namely the coordinates x and y of the plane, at the core of 2D scatter plot representation.
- Form, including size, shape (qualitative only) and orientation.
- Colour, with the three components of the colour space (as detailed in Sect. 1.3.5).
- Texture, including spacing, granularity and pattern (qualitative only).
- Optics, considering blur, transparency, shading and stereoscopic depth z in 3D scatter plots.
- Motion, including speed and rhythm by adding a temporal dimension to static images.

1.3.2   Spatialization and Gestalt Principles

Spatialization [139] consists in representing abstract metric data spatially. In other words, it relies on the embedding of metric data in a low dimensional space that may be visualized, often referred to as a map. The proximity and remoteness between the points in the map should reflect as faithfully as possible the similarities and dissimilarities between the associated data points. The visual encoding of a map is most often based on scatter plots.

Based on the map, data analysts attempt to identify structures present in the data space. However, those structures may differ between the data and the map due to distortions by the dimensionality reduction step. Hence, it is only possible to identify map structure, such as map clusters or map outliers. As such it is necessary to verify that those structures actually correspond to effective data structures, namely associated data clusters and data outliers (see Chap. 4).

The particular case of visual clustering relies on the Gestalt principles of grouping [199]. Those principles describe how humans visually perceive groups of items (with Gestalt meaning pattern or shape in German). This perception may be pre-attentive [139], meaning that pattern information can be processed without conscious cognition, providing information with minimal cognitive load. Visual exploration with low-dimensional scatter plots mainly relies on the following principles:

- *Proximity*: points that are placed close to each other in the scatter plot will be intuitively grouped together. This principle is at the core of many dimensionality reduction techniques trying to preserve distances in the mapping process.
- *Continuity*: collections of points forming connected dense parts of the space are detected as groups, even if points from the group might be closer to points from another group than to points from their group. This allows to perform visually some sort of density-based clustering.
- *Similarity*: points with similar markers, for example in terms of shape or colour, are identified as part of the same group. This is traditionally used to convey information about classes or data clusters.

- *Closure*: missing parts of virtual closed lines forming the contours of a given shape may be guessed with the brain intuitively filling the blanks. This principle might also be used to present classes or data clusters through the winglets method [120]. This method adds small wing-like lines locally following the density contours of the class, allowing to distinguish efficiently groupings of points despite the overlap of groups in the map. This may be combined with colour, to also benefit from the similarity principle.

1.3.3 Scatter Plots

Scatter plots represent data as markers in a Cartesian grid, with axes corresponding to each coordinate of the considered space. They are the common way of representing low-dimensional data, such as obtained with Dimensionality Reduction (DR).

1.3.3.1 2D and Interactive 3D Scatter Plots

Two-dimensional scatter plots allow to display points in a 2D space. With this encoding, the main information is conveyed through position which is a very reliable perceptual channel [17]. This type of encoding extends to sets of points in a three-dimensional space with interactive 3D scatter plots, which use stereoscopic rendering of the third dimension and may be rotated to observe the map from different angles. However, they suffer from several drawbacks such as occlusion of objects, ambiguity in depth perception, distortion of distances and angles due to perspective, and difficulty to interact and navigate the figure [163]. Scatter-plots with many points may also be susceptible to overdraw problems, with close points markers drawn on top of each other, thus hiding part of the points when many are regrouped in a small portion of space. This may be dealt with by representing points by their distribution (obtained by kernel density estimation) in the densest regions, allowing, for example, to represent more clearly overlaps between groups of points [124].

1.3.3.2 Circular Background and Procrustes Transform

When representing a map obtained through Dimensionality Reduction (DR) with a 2D scatter plot, domain experts, which are often unfamiliar with DR, might be confused by the axes of the map. Indeed, scientists confronted with a chart are often trained to search for which variable corresponds to each axis, or what value range is represented. Yet, in most non-linear dimensionality reduction methods, the synthetic axes have no clear meaning and only the distances between points matter. As a result, the map may be rotated, symmetrized, translated or even scaled (provided that distances values do not convey much meaning for the user), without impacting the information that it conveys. To remove this source of confusion and underline

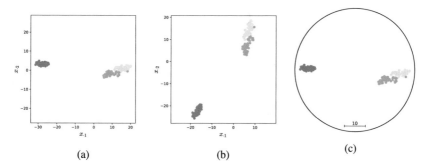

Fig. 1.13 Two equivalent maps of the iris dataset obtained with the DR method tSNE, displayed both with axes and circular background. Axes wrongly imply that the orientation of the map is meaningful. (**a**) Map 1 with axes. (**b**) Map 2 with axes. (**c**) Map 1 with circular background

invariance to these transformations, maps may be represented without axes and with a circular background [53]. We may note however, that for some methods such as PCA, presenting the axes may remain meaningful.

Considering this invariance to the aforementioned operations, the visual comparison of several maps of the same dataset may benefit from alignment through the Procrustes transformation [21, 71]. This allows to find for a given map the combination of rotation, symmetry, translation and uniform scaling, leading to best fit a reference map in the least squares sense (Fig. 1.13).

1.3.3.3 Glyphs

The shape of markers int a 2D scatter plot may be used to map several variables at a time using symbols called glyphs, such as flower glyphs or star plots [95]. A glyph provides a radial display in which each variable is represented along a different axis identified based on its angular orientation (as in radar chart).

1.3.3.4 Scatter Plot Matrices (SPLOM)

For representing points in more than three dimensions, Scatter PLOt Matrices (SPLOM) [178] proposes to combine many 2D scatter plots representing every possible pairs of variables plotted against each other (see Fig. 1.14). These plots are organized in a matrix, with the chart in the ith row and the jth column showing the variable $x_{\bullet i}$ plotted against variable $x_{\bullet j}$. Those individual scatter plots correspond to orthogonal projections on the subspace defined by the selected axes i and j. The diagonal charts (i, i) can be filled with the histogram of variable i. SPLOM representation is analogous to the covariance matrix, since each view of the SPLOM and each element of the covariance matrix characterizes the co-evolution of two variables [200]. The two mainly differ by their level of aggregation.

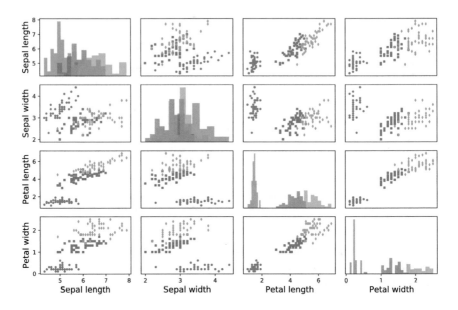

Fig. 1.14 Example of SPLOM for the iris dataset

Such representations enable analysts to visualize embedding of high dimensional data into spaces of dimensionality $d \geqslant 4$. However, as the number of variables to represent increases, the time cost of consulting all scatter plots explodes, with around d^2 different visualization to consider successively, as does the complexity of constructing a mental image of the data based on its representation [163].

To reduce the number of scatter plots that the analyst needs to consult, some indicators have been conceived to account for their individual usefulness. Hence, it is possible to filter out the less informative views. These indicators, called Scagnostics (for scatter plot diagnostics) [178, 200], provide a quantitative evaluation of several qualitative attributes of scatter plots. For example, a scatter plot may be qualified as monotonic, straight, skinny, stringy, striated, convex, clumpy, skewed or outlying [200]. The variety of those indicators may help identify the most appropriate view depending on the task at hand. For example, a more outlying view may be more appropriate for outlier detections, while a monotonic view may reveal correlations between variables. Some other indicators have been designed in this mindset, such as class consistency [168], taking into account the separability of predefined classes.

1.3.3.5   Grand Tour

The grand tour [8] shows a sequence of 2D views of a high dimensional dataset, obtained by successive orthogonal projections. These views are presented as frames of a film. To reduce the required cognitive load, two consecutive views must be relatively close, inducing a continuous motion around the data space. Hence, the sequence of views is chosen so as to be dense in the space of all possible views. This approach relates to the SPLOM, but spreads the different view over time instead of over space. Similarly to SPLOM, the sequence may be selected to be composed of more informative views based on specific criteria [40].

1.3.4   Parallel Coordinates

Parallel coordinates [92] provide a visual representation of high dimensional data that does not require dimensionality reduction (see Fig. 1.15). All dimensions of the data space are represented by independent vertical axes, that may have their own scales, and that are distributed regularly along the horizontal axis. Then, one line chart is drawn for every high dimensional point, crossing each vertical axis at the value of the associated coordinate for the point. Thus, close points in the data space are represented for bundles of curves in this representation, so that parallel coordinates may be used for visual clustering: [210].

The order of dimensions may significantly impact the interpretation of the visualization. Thus, automatic ordering techniques may be used to select the best arrangement, for instance by placing close together other dimensions that are similar in some sense [7]. These similarities may be measured, for example, using distances on the normalized column vectors of the data matrix.

For sequence data treated as multidimensional data (see Sect. 1.1.5), the parallel coordinates representation (with uniform vertical axis) simply corresponds to plotting all the sequences against their common index.

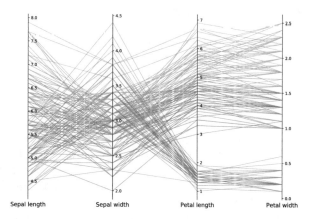

Fig. 1.15   Example of parallel coordinates for the iris dataset

1.3.5   Colour Coding

Much information may be encoded through colour, relying on a colour map linking each different value of a variable to a specific colour.

1.3.5.1   Colour Models

Colours may be considered as points in a three dimensional (3D) space. Due to low-level biological (treatment of colour both by the cone cells of the human eyes as a tri-stimulus) and hardware reasons [127] (display by monitor screens), they are often defined as a combination of red, green and blue components, leading to the RGB colour model. However, many other colour models have been constructed, which are simply different coordinate systems for the same colour space.

The HSV colour model decomposes colours into three more qualitative descriptors, which are the hue (tint), saturation (colour intensity) and value (colour lightness). Other colour models, called perceptual, have been defined based on empirical studies on effective perception of colours. This is the case of CIELAB or CIELUV [39] defined by the "Comission Internationale de l'Éclairage" (CIE), that both use in their coordinate system the luminance L^* of colours. In these perceptual models, the Euclidean distance between the coordinates of two colours directly measures the perceptual difference between those colours. Perceptual considerations focus on a standard observer. However, colour perception is not universal and varies across people. In addition, some individuals suffer from colour-blindness. This corresponds to a deficiency in the perception of certain colours, leading to a restriction of the perceived colour space.

1.3.5.2   Taxonomy of Colour Maps

Variables may be visually encoded through colours, linking every value to a specific colour based on a colour map. We may distinguish colour maps between *colour palettes* used to represent qualitative (or categorical) variables and *colour ramps* used to represent quantitative (ordinal or numerical) variables [169]. Those different types of colour maps are illustrated Fig. 1.16.

For a categorical variable with n categories, a good colour palette mainly needs to provide n discriminable colours. A secondary requirement is that it satisfies certain criteria of aesthetics, which is obviously subjective. In the case of categories strongly associated with colours in the viewers mind, for example due to cultural reasons, the choice of a colour palette may also be subject to concerns about colour-concept associations [150].

In addition to colour discriminability and aesthetics, colour ramps need to provide a natural ordering and a uniform perceptual gradient. The rainbow (or jet) colour map, ranging from blue to red with a linear change of hue (in the order of

Fig. 1.16 Different types of colour maps

wavelengths of visible light spectrum), does not fit either of these requirements. Yet, it is still set as default by some visualization software and widely used in scientific publications, causing much rage in the visualization community [22, 152]. However, these may be obtained by regular variations of luminance or of saturation. Empirical studies [152] show that luminance-based colour ramps are better-suited for mapping variables that change with high spatial frequency, while saturation-based (isoluminant) colour ramps are more appropriate for variables with low spatial frequency. Isoluminance may also be an interesting property for use with three-dimensional (3D) representations, luminance being perceptually linked with stereoscopic depth [22].

Sequential colour ramps, ranging continuously from a minimum to a maximum value, are often encoded with a unique hue and a monotonic change of luminance or saturation. Oppositely, diverging colour ramps, corresponding to symmetric evolution around a central value (for example zero, or the mean of the variable) may use two opposite hues on both sides of the central value with monotonic evolution of luminance or saturation when moving away from the central value.

Many off-the-shelf colour maps are available that have been hand-crafted by designers, with good perceptual properties, some of them colour-blind safe (adapted for colour-blind viewers) [27, 74]. Yet, some tools are developped to help naive users to automatically construct good personalized colour maps suiting their specific needs [150, 169]. Finally, some bi-dimensional and tri-dimensional colour schemes may be used to represent thet co evolution of two or even three variables [74, 112].

1.3.6 Multiple Coordinated Views and Visual Interaction

Visual exploration of a dataset does not need to rely on a single representation, and multiple views may be used at once. Such dashboard representations with different panels are used for example in [95, 132, 170]. The views may correspond to the same type of visual encoding representing different features of the data (as in SPLOM or the two graphs view of MING), or representations of different types. The latter may, for instance, combine a 2D scatter plot of a map obtained with dimensionality reduction and a parallel coordinates representation of the high dimensional data.

Moreover, views do not need to be static, and users may interact dynamically with them. User interaction relies on operations such as filtering and highlighting (or brushing), to focus on specific elements of the dataset, or zooming and panning for moving in the coordinates system of the figure axes.

When using multiple views alongside with interaction, certain principles, such as data linking and navigational slaving may help to reduce users confusion in context switching and in comparison of different views [195]. Data linking relates the individual data points to their counterparts in the other views, so that filtering and highlighting affect all instances of a given data point simultaneously across all views. It may be complemented by visual cues helping users to identify that those instances represent the same data point, for example by using a similar colour and/or marker. Navigational slaving relates all axes corresponding to a similar feature, even between different types of representations (e.g., a scatter plot and a bar chart), so that zooming and panning operation allows to move simultaneously along the axes common to several views. For example, in a SPLOM, navigating the figure (i, j) would have repercussions on all figures of row i and column j. Navigational slaving is also used between the two figures of MING.

Zoomingt and panning may provide a better view of some part of the data, but in doing so it erases the big picture, so that the focus information becomes out of context. This may be alleviated by using either an *overview+detail* or a *focus+context* approach [132].

1.3.7   Graph Drawing

Network data are often represented as node-link diagrams [101]. The vertices are represented by points whose positions are obtained by graph embedding (or graph layout) methods, which is the equivalent of DR for network data. The edges are shown as lines between those points. This representation may be subject to visual clutter due to multiple edges crossings and overlaps. To alleviate that problem, one may use edge bundling, to agglomerate similar edges together [89, 91, 146, 165, 212].

1.4   Intermediate Conclusions

DR is a tool for extracting quantitative variables describing the variability in metric data, which encompass multidimensional, sequence and network data. As such, it may be related to other tools of data science, such as clustering, classification and regression. DR and clustering allow to identify categories or latent variables in a dataset based on its structure. Conversely, classification and regression estimate the target variable (i.e. category or value) associated to a data point by inferring the relation between the data features and the target variable from training points. For visual exploration, DR may be used to get a spatial representation of metric data usually displayed with a scatter plot.

Chapter 2
Intrinsic Dimensionality

The search for truth is more precious than its possession.

Albert Einstein

High dimensional data are subject to the curse of dimensionality defined in Sect. 2.1, which hinders their analysis. Yet, in practice, data may be assumed to live in a manifold whose dimensionality is lower than that of the data space dimensionality. The effective dimensionality of data is called intrinsic dimensionality. Its estimation is detailed Sect. 2.2.

2.1 Curse of Dimensionality

The curse of dimensionality [1, 57] is a broad term encompassing several phenomena occurring in high dimensional spaces, and hindering the capability to handle very high dimensional data.

2.1.1 Data Sparsity

One of the main issues with high dimensional spaces is the sparsity of data [57]. Figure 2.1 illustrates this problem by considering the tessellation of a hypercube with a Cartesian grid. We see that this requires a number of points increasing exponentially with the dimensionality. Indeed, to regularly sample each of the δ dimensions with n_0 regularly spaced points necessitates a total of n_0^d points. Thus, a tractable number of data points is necessarily insufficient to cover the entire data space.

S. Lespinats et al., *Nonlinear Dimensionality Reduction Techniques*,
https://doi.org/10.1007/978-3-030-81026-9_2

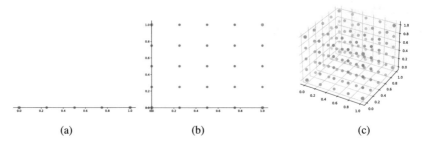

Fig. 2.1 Exponential increase of the number of points required to regularly tesselate an hypercube. (**a**) 1D. (**b**) 2D. (**c**) 3D

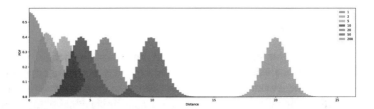

Fig. 2.2 Distribution of Euclidean distances between points randomly drawn from the same Gaussian distribution with unit variance

2.1.2 Norm Concentration

The phenomenon of norm concentration (or concentration of measure) impacts the discriminating power of dissimilarity measures in high dimensional spaces. Figure 2.2 illustrates this phenomenon with the distributions, for several values of the dimensionality δ, of Euclidean distances between points drawn from a common δ-dimensional Gaussian distribution with unit variance. In this case, the distribution of distances is a χ distribution with δ degrees of freedom [107]. Thus, when $\delta \to \infty$, its mean behaves asymptotically as $\sqrt{\delta}$, while its variance tends towards a constant, resulting in a loss of relative contrast between the small and high distances when the dimensionality increases. This evolution of distances distribution also leads to an under-sampling of small distances for high dimensional datasets [81]. This phenomenon has also been studied for more general data distributions and the broader family of Minkowski distances (presented Sect. 1.1.4). Considering the contrast measure $\frac{\max_{k \neq i} \Delta_{ik} - \min_{k \neq i} \Delta_{ik}}{\min_{k \neq i} \Delta_{ik}}$, for distances from a point i, Minkowski distances have been shown to be more susceptible to norm concentration when their parameter p increases [1].

Despite the effects of the curse of dimensionality, analysis of data in high dimensional spaces remain conceivable. remains possible to consider applications in high dimensional spaces. This is based on the assumption that data live in a manifold whose intrinsic dimensionality is far lower than the ambient space

Fig. 2.3 Effective distribution of Euclidean distances between points of the handwritten digits dataset. These are separated between intra-class and inter-class distances, and the two distributions are normalized accordingly

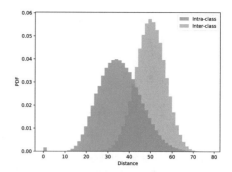

dimensionality. This hypothesis may be justified by the existence of relations between the considered variables, thus reducing the effective number of degrees of freedom [45]. Furthermore, for real datasets, norm concentration does not necessarily induce a critical loss of discriminability for distances. Considering for instance the case of the *digits* dataset in a 64−dimensional space, we notice that high and low distances are not completely confused (*cf.* Fig. 2.3).

2.2   Estimating Intrinsic Dimensionality

The intrinsict dimensionality ∂ corresponds to the dimensionality of the data manifold $\mathcal{M}$ and is a key parameter characterizing the complexity of a dataset. Yet, defining and estimating intrinsic dimensionality remains a complex task. For a linear subspace, its dimensionality corresponds to the number of vectors required to form a basis of this subspace, that is a family of linearly independent vectors generating the subspace. In other words, it indicates the number of coordinates allowing to define all the points as linear combinations of common basis vectors. This definition characterizes the dimensionality δ of the ambient data space, as well as the dimensionality d of the embedding space. Similarly, the intrinsic dimensionality of data intuitively represents the number of latent variables required to parametrize locally the data manifold, namely the degrees of freedom in that manifold. It can be defined more formally as the dimensionality of the linear tangent subspace at each point [81]. As such, it is a local property and may be defined for each data point as ∂_i. Figure 2.4 represents such tangent subspace for a one-dimensional and two-dimensional manifolds, respectively in a 2D and 3D data space. The current Section presents practical approaches for estimating the intrinsic dimensionality associated to a discrete dataset.

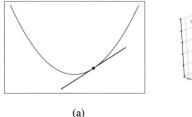
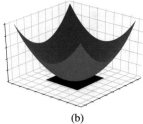

(a)   (b)

Fig. 2.4 Tangent spaces (black) for manifolds (blue) of different intrinsic dimensionality. For the one-dimensional parabola manifold immersed in a two-dimensional data space, the tangent space is a line, while for the two-dimensional paraboloid manifold, in a three-dimensional data space, the tangent space is a plane. (**a**) 1D. (**b**) 2D

2.2.1   Covariance-Based Approaches

Covariance-based approaches attempt to determine the number of components describing meaningful variability in the data, while discarding variability associated to noise.

2.2.1.1   Scree Plot

Assuming a linear manifold (morally, we are speaking of a vector space), the scree-plot method [32] evaluates the intrinsic dimensionality as the minimum embedding dimensionality for which Principal Component Analysis (**PCA**) [94, 141] retains a sufficient amount of the data variance. Indeed, **PCA** identifies principal components defining a set of decorrelated synthetic variables, and obtains the embedding by selecting the variables with the highest variance among that set. The scree-plot, illustrated Fig. 2.5, presents the amount of variance associated to each of those principal components as a function of their index (ordered by decreasing variance).

The analyst can then identify an "elbow" in that curve (which is subjective), that is an index for which there is a sudden change of the curve slope (see the blue point in Fig. 2.5). This rule of thumb provides a threshold to discriminate between principal components corresponding to degrees of freedom of the manifold (with significant variance) and principal components corresponding to high dimensional noise (residual variance) separating the components that correspond to a significant amount of the total variance from those that correspond to a residual amount. Hence, the intrinsic dimensionality is given by the index of that "elbow", which indicates the number of principal components with significant variance. We may note that the determination of the elbow is subjective, but the same could be said of a user-defined quantitative threshold.

Furthermore, for a non-linear manifold, the scree plot dimensionality indicates the dimensionality of the linear subspace containing the manifold. Considering that

Fig. 2.5 Scree plot for the 12−dimensional Wine dataset (black curve). A visually defined elbow in the curve is located by a square blue marker. The estimation of that dimensionality with a fixed threshold $1 - \epsilon$ of the cumulated proportion of the total variance set to 95% is also illustrated by the gray curve and orange lines

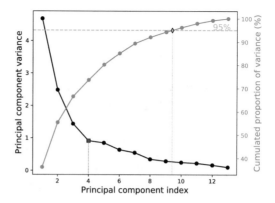

the manifold is folded in that subspace, we call this the *folding dimensionality* which would be an upper bound for the intrinsic dimensionality (i.e., the effective number of degrees of freedom). For instance, a sphere has an intrinsic dimensionality of two but a folding dimensionality of three.

2.2.1.2 Local Covariance Dimension

The local covariancet dimension [45] relies on a similar approach to estimate locally the intrinsic dimensionality ∂_i, considering that the manifold is locally linear. For a given point i, the covariance matrix is computed considering only the points in a given neighbourhood of the point i. Since the scree plot analysis may not be manually used for each point i, a criterion is set to automatically determine the low variance components to be discarded. This criterion is defined as follows: for a given $\epsilon \in [0, 1]$, corresponding to the proportion of the total variance presumably explained by noise, the dimensionality is set to the minimum value ∂_i satisfying:

$$\frac{\sum_{k=1}^{\partial_i} \lambda_k}{\sum_{k=1}^{\delta} \lambda_k} \geqslant 1 - \epsilon. \tag{2.1}$$

In the above Equation, λ_k, with $k \in [\![1; \delta]\!]$, are the eigenvalues of the covariance matrix, namely the variances associated to each principal component, sorted in decreasing order. Figure 2.5 also represents a slight adaptation of this automated rule for the covariance matrix of the entire **Wine** dataset. This adaptation extends the local covariance dimension to non-integer dimensionality values by linear interpolation of the ratio in Eq. (2.1), considering the value of δ_i for which the equality case is reached. We may note that this measure is upper-bounded by κ, the number of points other than i in the considered neighbourhood, since $\kappa + 1$ points can only define a linear space of dimensionality κ. Thus, the neighbourhood size must be chosen accordingly.

2.2.2 Fractal Approaches

An alternative definition of dimensionality comes from the field of fractals, allowing fractional dimensionality values for fractal objects such as the Cantor set or Koch snow flake [121]. In broad terms, the fractal dimension ∂ of an object is defined according to a scaling law of the form $\mu \propto \sigma^{\partial}$, linking some measure μ of the volume of a small element of that object space to the scale σ of that element. The volume accounts for the portion of space filled by the element. As such, several definitions, such as Hausdorff (or box-counting) dimension [82] and Assouad dimension [45], consider the evolution of the number of base elements under a certain diameter required to cover the space (or more precisely the cumulated diameter of those elements), which is inversely proportional to the volume of those elements. Yet, for high dimensional datasets, determining a cover of the manifold is impractical, so that the correlation dimension, which yields close results in many cases, is used instead [82].

2.2.2.1 Correlation Dimension

The correlationt dimension [29, 82] relies on the assumption that a manifold of intrinsic dimensionality ∂ locally behaves as a Euclidean space of that dimensionality. In such a Euclidean space, the volume of a ball of radius σ is proportional to σ^{∂}. Hence, considering that the density of points in the manifold is uniform, the number of neighbours in the neighbourhood of size of any given point is proportional to that volume for sufficiently small values of σ. As a result, for a set of N points uniformly sampled from that manifold, the proportion of point pairs within a distance σ from each other, called $C(\sigma)$, follows for $\sigma \to 0$, the scaling law:

$$C(\sigma) \propto \sigma^{\partial} \quad \Longleftrightarrow \quad \log C(\sigma) = \partial \log \sigma + \text{const.} \tag{2.2}$$

In practice, the correlation dimension is obtained as the slope of the linear part of the graph of $\log C(\sigma)$ against $\log \sigma$ [82], as illustrated in Fig. 2.6a.

However, this approach is subject to a conflicting requirement for the definition of the linear part. The scales σ must be small enough to satisfy the scaling law of Eq. (2.2), while the largest of those scales must be sufficiently high to encompass a statistically significant number of pairwise distances [59]. Moreover, due to the curse of dimensionality, the proportion of small distances decreases with the dimensionality of the manifold, so that very big datasets are necessary to reach the same quality of estimation. As such, for high dimensional manifolds with a reasonable number of samples the correlation dimension leads to a systematic under-estimation of the dimension. This is confirmed by Fig. 2.6b, showing that the estimated slope decrease with the number of samples.

To tackle this issue, several solutions have been proposed. An empirical method [29] consists in estimating the connection between the estimated dimensionality

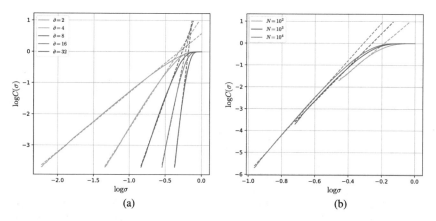

Fig. 2.6 Correlation dimension curves for points uniformly sampled from unit hyper-cubes. The variation with dimensionality (left panel) and number of samples in the dataset (right panel) are shown. (**a**) Dependency to dimensionality ($N = 2000$). (**b**) Dependency to number of samples ($\partial = 10$)

and the true dimensionality based on synthetic datasets of known dimensionality and containing the same number of samples as the dataset of interest. This relation may then be inverted to obtain a corrected estimation of the intrinsic dimensionality from the biased estimate of the dimensionality. Another possibility is to consider the geodesic distances (approximated with shortest path distances on a nearest-neighbours graph as detailed in [175]), instead of the ambient space distances [81]. As such, the manifold curvature is accounted for, allowing to extend the range of values σ satisfying the scaling law to higher scales. This also enables to define a scale dependent dimensionality by considering the slope at each point of the curve $\log C(\sigma)$ against $\log \sigma$.

2.2.2.2 Nearest Neighbours Dimension

The two-NN estimator [62] considers points drawn from a locally uniform distribution and that the probability to find a point in a region of space is proportional to its volume. With those assumptions it may be shown that the ratio η between the distance to the second and first neighbours of a point follows the Pareto distribution with parameter ∂. Compared with the correlation dimension, this approach allows to drop the hypothesis of a uniform density over the entire dataset.

The Paretot distribution corresponds to the probability density function f and the associated cumulative density function F (illustrated Fig. 2.7a):

$$f(\eta) = \partial \eta^{-\partial-1} 1_{[1,+\infty[}(\eta) \quad \text{and} \quad F(\eta) = (1 - \eta^{-\partial}) 1_{[1,+\infty[}(\eta) \tag{2.3}$$

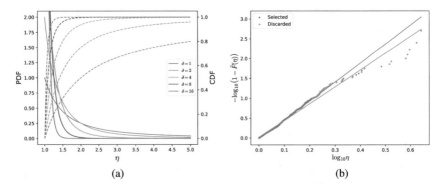

Fig. 2.7 Pareto distribution for different values of the parameter ∂ (left panel) and estimation of the intrinsic dimensionality of a Cauchy distributed (many outliers) five-dimensional dataset with the two-NN method [62]. The orange points indicate the discarded 10% of points. The blue line is fitted to the blue points only, while the orange line is fitted to the blue and orange points. (**a**) Pareto distribution. (**b**) Estimation with two-NN method

where $1_{[1,+\infty[}$ is the indicator function of the set $[1, +\infty[$ and the ratio of distances is formally defined for each point i as:

$$\eta_i \triangleq \frac{\Delta_{i\tilde{v}_i(2)}}{\Delta_{i\tilde{v}_i(1)}}, \tag{2.4}$$

with $\tilde{v}_i$ the neighbourhood permutation for point i (as defined in Sect. 1.1.2). Considering the expression of the Cumulative Distribution Function (CDF) provided by Eq. (2.3), the parameter ∂ of that distribution may be estimated as the slope of $-\log\left(1 - \widehat{F}(\eta)\right)$ against $\log \eta$, where $\widehat{F}(\eta)$ is the empirical estimation of the CDF. To obtain this slope, the two-NN method fits a line to that curve by linear regression removing the 10% of the points with highest ratio η. This filtering allows to alleviate the effect of outliers which imply high variations of local density, thus violating the hypothesis. Figure 2.7b illustrates this method for a dataset with strongly varying density.

2.2.3 Towards Local Estimation

For the abovet fractal dimensions estimators, the dimensionality is considered constant for the entire dataset. Yet, it is a local property and may vary along the manifold.

Figure 2.8 presents the estimation by the two-NN method of the intrinsic dimensionality for three datasets. The first and second datasets are each constituted of one Gaussian cluster of a given dimensionality, while the third is constituted of

Fig. 2.8 Estimation of the dimensionality of a 3D and 7D Gaussian, as well as a dataset constituted of two well-separated Gaussian clusters (respectively of 3D and 7D)

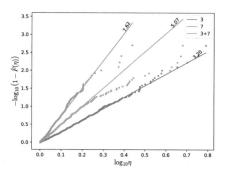

two well-separated Gaussian clusters of different dimensionality. In the latter case, the estimated dimensionality is intermediate between those of the two clusters.

2.2.3.1 Hidalgo

Based on the two-NN method,t the Heterogeneous Intrinsic Dimension Algorithm (Hidalgo) [3] proposes a local estimation of the intrinsic dimensionality. It assumes that the dataset is constituted of K manifolds each of a given dimensionality ∂_k. Based on this hypothesis, it models the distribution of the ratio η_i by a mixture of Pareto distribution, whose Probability Density Function is of the form:

$$p(\eta_i) = \sum_{k=1}^{K} w_k \partial_k \eta_i^{-\partial_k+1} \quad \text{with} \quad \sum_{k=1}^{K} w_k = 1,$$

where each component k has a weight w_k. A Bayesian modelling approach is then used to determine the dimensionality and weights of the components, and for each data point, the component to which it belongs. Yet, these Pareto components being largely overlapping (as shown Fig. 2.7a), a spatial dependency is incorporated. This allows to account for the fact that close-by data points are more likely to belong to the same component. Finally, the parameters of the Bayesian model are obtained by Gibbs sampling.

2.2.3.2 Hill Estimator

Hill estimator [6] gives a local estimation of a continuously varying intrinsic dimensionality based on extreme value theory. Considering for point i a neighbourhood of radius ω and cardinal κ, it estimates the intrinsic dimensionality in i as:

$$\widehat{\partial}_i^{\text{Hill}} = \left(\frac{1}{\kappa} \sum_{j \in v_i(\kappa)} \log\left(\frac{\omega}{\Delta_{ij}} \right) \right)^{-1}. \tag{2.5}$$

When κ is tends to ∞ this estimator is shown to converge in distribution toward a normal distribution $\mathcal{N}\left(\partial_i, \frac{\partial_i^2}{\kappa}\right)$ using the Central Limit Theorem (CLT).

2.3   Two-Nearest Neigbhours Intrinsic Dimensionality Local Estimator

We introduce here a new approach derived from the two-NN estimator for estimating locally the intrinsic dimensionality. We call this method Two-nearest neighbours Intrinsic Dimensionality Local Estimator (TIDLE).

2.3.1   Gaussian Mixture Modelling

This estimatort assumes that the variation of the local intrinsic dimensionality ∂_i is of low spatial frequency, so that ∂_i may be considered constant over a sufficiently small neighbourhood. Hence, the ratio η_i defined for the two-NN method (Eq. (2.4)) follows a Pareto distribution with parameter ∂_i in the neighbourhood of a point i. The maximum likelihood estimate of that parameter is given by [137]:

$$
\widehat{\partial_i} = \left(\frac{1}{\kappa} \sum_{j \in v_i(\kappa)} \log \eta_j\right)^{-1},
\tag{2.6}
$$

where $v_i(\kappa)$ is the κ-neighbourhood of point i. Formally, the estimator of Eq. (2.6) computes the geometric mean of the ratios η_j over the neighbourhood of i, and then take the inverse of its logarithm. Hence, this geometric mean tends to alleviate the impact of the high ratios η_j which are discarded in the case of the two-NN method.

For j in a neighbourhood of i, η_j is drawn from a Pareto distribution with parameter ∂_i. Hence, $\log \eta_j$ is drawn from an exponential distribution with parameter ∂_i (see Annex A.2). The mean and variance of that random variable are:

$$
\mathrm{E}[\log \eta] = \frac{1}{\partial_i} \quad \text{and} \quad \mathrm{Var}[\log \eta] = \frac{1}{\partial_i^2}.
$$

Samples $\log \eta_j$ being independent and identically distributed, the Central Limit Theorem (CLT) indicates that their arithmetic mean (i.e. $\widehat{\partial_i}^{-1}$) converges in distribution towards a Gaussian distribution $\mathcal{N}(\mathrm{E}[\log \eta], \frac{\mathrm{Var}[\log \eta]}{\kappa})$. Assuming that the dataset is constituted of a few components each of uniform dimensionality the distribution of the $\widehat{\partial_i}^{-1}$ may reasonably be modelled by a simple Gaussian mixture

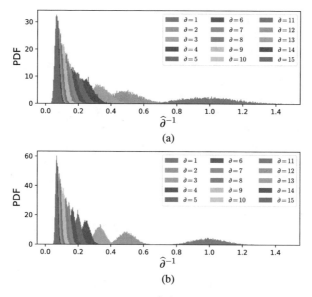

Fig. 2.9 Distribution of the inverted estimator $\widehat{\partial}_i^{-1}$ for different values of the intrinsic dimensionality ∂_i and different neighbourhood sizes κ. This is obtained by generating η_i with a Pareto distribution with parameter ∂_i. (**a**) Distribution for $\kappa = 30$. (**b**) Distribution for $\kappa = 100$

[126]. Namely, their probability density function is given by:

$$p(\widehat{\partial}_i^{-1}) = \sum_{k=1}^{K} w_k \, p_k(\widehat{\partial}_i^{-1}), \quad \text{with} \quad \sum_{k=1}^{K} w_k = 1.$$

In this Equation, K is the number of components, w_k is the weight associated to the kth Gaussian component and p_k is the Probability Density Function (PDF) of that component.

Figure 2.9 illustrates the components of that Gaussian mixture, that is the empirical distributions of $\widehat{\partial}_i^{-1}$ for neighbourhoods of size κ with uniform dimensionality ∂_i. We see that proper discrimination of two components requires sufficiently high values of κ, especially for components associated to high ∂_i. Yet, this value is bounded from above by the necessity to have a locally uniform intrinsic dimensionality.

The optimal number of components for that mixture is determined by minimizing the Bayesian Information Criterion (BIC) [126]. This criterion, used for model selection, measures the negative log-likelihood of the model for a set of N samples $\widehat{\partial}_i^{-1}$. It also penalizes the number of components K to avoid over-fitting, considering that the likelihood naturally increases with the number of components. It is defined as:

$$\mathrm{BIC} \triangleq K \log N - \sum_{i=1}^{N} \log p(\widehat{\partial}_i^{-1}).$$

A possible solution is to use standard Gaussian mixture modelling with K Gaussian components of unknown mean and variance, whose likelihood is maximized by the Expectation Maximization (EM) algorithm. Replicating this for $K \in [\![1, K_{\max}]\!]$ the best number of components can then be obtained by minimizing the BIC.

We consider here an alternative solution for which the dimensionality parameters are assumed to be integers. Hence, the Gaussian component k associated to $\partial = k$ has known mean and variance and correspond to $\mathcal{N}(\frac{1}{k}, \frac{1}{\kappa k^2})$. In that case, we may optimize the BIC with respect to the weights $\{w_k\}$. Many of the weights $\{w_k\}$ being close to 0, the effective number of components used to compute the BIC is lower than K. We may thus assess it with the perplexity $\widehat{K}$ of the weights distribution, which has the advantage of being a differentiable function of the weights:

$$\widehat{K}(\{w_k\}) \triangleq \exp\left(-\sum_{k=1}^{K} w_k \log w_k\right).$$

TIDLE provides an estimate of the distribution of dimensionality in a dataset. It can also be used to determine the point-wise intrinsic dimensionality either by assigning each point to the component most likely to explain its ratio η_i. A more local estimation can be determined as the average of the dimensionality of all components weighted by the probability that those components explain the ratio η_i.

2.3.2   Test of *TIDLE* on a Two Clusters Case

To test our approach,t we consider simple synthetic datasets, each constituted of two clusters of different dimensionality generated by a multivariate distribution restricted to a linear subspace of the appropriate dimensionality in a 10−dimensional ambient space. These clusters are randomly rotated [55], and translated along the diagonal $(1, 1, \ldots, 1)$ of the unit hypercube so as to ensure a given distance between their centroids. An example of a similarly generated dataset in a three-dimensional space is shown in Fig. 2.10a. Figure 2.10 presents the weights w_k identified by TIDLE for the mixture as a function of the component index k. The variability of the method is accounted for by considering in each case 100 datasets randomly generated with the same characteristics. The distribution of each weight w_k over the 100 runs is shown as a box plot. For all datasets, the clusters each contain 500 points and the neighbourhood size is set at $\kappa = 100$.

Figure 2.10b presents the results for the standard case, where the dimensionality of the two Gaussian distributed clusters is respectively three and seven. Those clusters (of unit variance) are well-separated with a distance of five hypercube diagonals, and subject to no noise. The three-dimensional component is detected in at least 95% of the runs and in these cases its associated weight is higher than 0.3, as shown by the fifth percentile of the distribution of w_3. As for the seven-

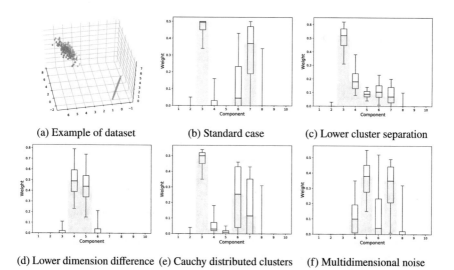

(a) Example of dataset (b) Standard case (c) Lower cluster separation

(d) Lower dimension difference (e) Cauchy distributed clusters (f) Multidimensional noise

Fig. 2.10 Boxplots of the weights w_k obtained by TIDLE as a function of the index k of the component for 100 randomly generated two clusters datasets. Panel (**a**) presents an example of two clusters dataset with one- and two-dimensional clusters in an ambient three-dimensional space. Other panels each correspond to a specific set of characteristics for the generation of the datasets. Panel (**b**) corresponds to a standard case with well-separated three- and seven-dimensional Gaussian distributed clusters in a ten-dimensional ambient space. Panel (**c**) to (**d**) show the impact on the estimator of changing some characteristics of this standard case

dimensional component, it is detected in at least 75% of cases, considering the first quartile of the distribution of w_7. Yet, for some datasets, several components of close dimensionality are detected for a same cluster. In particular, the seven-dimensional cluster is also partly explained by a six-dimensional component in at least 50% of cases, considering the median of the distribution of w_6.

Figure 2.10c shows tthe effect of reducing the distance between the clusters of the standard case, using a distance only equal to half of the hypercube diagonal. Hence, the hypothesis of local uniformity of the density is hindered, which leads to the detection of many components of an intermediate dimensionality between three and seven.

Figure 2.10d corresponds to the case where the difference of dimensionality is lower, considering a four-dimensional and a five-dimensional cluster. In that case the variability of the weights is high for both components, but the two components are detected in more than 95% of cases. Thus, components of close dimensionality may be partly confused, but the variability of intrinsic dimensionality is correctly identified in most cases.

Figure 2.10e represents the case of Cauchy distributed clusters. This distribution implies high variations of density, so that the ratio η does not follow the Pareto distribution [62]. This leads to more underestimation of the dimensionality for the seven-dimensional cluster. This underestimation corresponds to the impact of outliers that may be observed Fig. 2.7b.

2.3.3   *TIDLE Perspectives*

TIDLE providest a local estimation of the intrinsic dimensionality assuming its local uniformity. It enables to identify the existence of several components of different dimensionality. Yet, it may also identify some artificial components, especially when the components of different dimensionality are overlapping. Future work should provide a thorough comparison of the results of TIDLE and Hidalgo for the local estimation of dimensionality, both in terms of accuracy and speed. We may note that, compared to Hidalgo, TIDLE proposes a simpler model with fewer parameters. Yet, it can not distinguish non-connected components of close intrinsic dimensionality and may be more sensitive to overlapping components of different intrinsic dimensionality. The difference between the estimator of Eq. (2.6) (estimating the inverse of δ) and Hill estimator (directly estimating δ) should also be investigated. Since both estimators follow a Gaussian distribution for a sufficiently high neighbourhood size, the Gaussian mixture approach of TIDLE could also be adapted with this indicator to detect specific components.

Chapter 3
Map Evaluation

> *A map is not the territory it represents, but, if correct, it has a similar structure to the territory, which accounts for its usefulness.*
>
> Science and Sanity
> Alfred Korzybski

The purpose of map evaluation is to assess the overall quality of a map. Most often, this quality is quantified by scalar indicators allowing to compare several maps. This may be used to select the best of several techniques for embedding a dataset or to tune the hyper-parameters of a given technique. It strives to measure the amount of distortions of the data structure, but differ from local evaluation for map interpretation (see Chap. 4), which aim at locating these distortions in the embedding space.

First, Sect. 3.1 discusses the challenge of evaluating the quality of a map. Then, Sect. 3.2 focuses on unsupervised indicators accounting only for the preservation of the data structure, while Sect. 3.3 details supervised indicators assessing the preservation of classes, based on additional class information.

3.1 Objective and Practical Indicators

Indicators are designed to assess the effectiveness of preservation of a specific type of information, sometimes in relation with a specific task that needs to be performed based on the map. Yet, the design of indicators to objectively assess DR results remains an open challenge.

3.1.1   Subjectivity of Indicators

Quantitative indicators for dimensionality reduction are by nature subjective [12]. Indeed, they rely on a loose definition of relevant data structure to maintain. This may also depend on the task that the user wants to perform with the output of the dimensionality reduction process. As opposed to supervised learning tasks, that may be evaluated with relative objectivity based on a classification or regression error on a validation set, defining quality indicators for dimensionality reduction involves several subjective choices. It requires to determine the information that needs to be retained, as well as a quantitative estimator assessing the preservation of that information. Moreover, those indicators are sometimes biased, since they are often built alongside with a dimensionality reduction method, considering the same paradigm of what a good map should be. Most DR methods relying on the minimization of a stress function, those stress function may even be used directly to evaluate the quality of maps [21, 189]. Yet, indicators designed specifically for the purpose of map evaluation are less constrained than DR stress functions by the smoothness or convexity requirements inherent to optimization (see Sect. 7.1). This allows, for instance, the use of neighbourhood ranks to assess map quality.

3.1.2   User Studies on Specific Tasks

Another way of assessing the quality of a map, or of comparing the quality of several maps is to perform a user studies. Such studies effectively measure whether a group of naive and/or expert users succeed in performing specific tasks using the considered map. Such an approach is, for example, used in [163] for the task of cluster verification. In this study, users rate for each dataset the effectiveness of data representation obtained with several dimensionality reduction techniques, but also with several values for the embedding dimensionality hyper parameters.

The user study approach reduces the bias induced by the ad-hoc modelling of human cognition necessary to define a quantitative indicator. Yet, it may introduce other biases due to the interaction with human subjects and the interpretation of their performance on a task. It is also more complicated to implement and may not be integrated in an automated pipeline. However, a mixed approach may integrate user studies in the design of quantitative indicators in order to verify which indicator best model human responses for a given task [11, 162].

3.2   Unsupervised Global Evaluation

Unsupervised indicators for map evaluation are used to assess the preservation of the data structure by the embedding, most often by penalizing a given type of distortion.

3.2.1 Types of Distortions

In the framework of the neighbourhood retrieval task [189], the goal affected to the mapping process is to embed points so that distances in the embedding space reflect dissimilarities in the data space. Thus, map evaluation needs to assess distortions of the neighbourhoods of points. These distortions occur at the elementary level of pairwise neighbourhood relations, which may be compared between the two spaces based either on distances or on neighbourhood ranks. To highlight the distinction between these metric and non-metric approaches, we use different terminologies to denote distortions in those distinct mindsets. Yet, in both cases, we may distinguish two main types of distortions: either points are represented too close from each other or they are represented too far.

Distortions can also be categorized as *mild* or *hard*. Mild distortions only impact the local geometry of data (local distance properties), while hard distortions affect the topology of the manifold (global properties). This distinction is mostly qualitative, but it may also be determined based on an arbitrary threshold (e.g., a neighbourhood size κ).

3.2.1.1 Distance Distortions

Distortions may be studied in terms of distances, mainly in the context of metric methods, which seek to preserve directly the values of distances. Perfect preservation occurs when $\Delta_{ij} = D_{ij}$ for all pairs of points (i, j). If the absolute values of distances is not deemed important (i.e. if their values are only considered relative to each other), distances may be normalized in both spaces, rather than comparing directly their values. For this metric consideration, we use the terminology introduced by [10]:

- when the distance between a pair of points is smaller in the embedding space than in the data space ($\Delta_{ij} > D_{ij}$), it is called a *manifold compression*. Hard manifold compressions are called *manifold gluings*.
- when the distance is larger in the embedding space than in the data space ($\Delta_{ij} < D_{ij}$), it is called a *manifold stretching*. Hard manifold stretching may be referred to as *manifold tears*.

3.2.1.2 Rank Distortions

Neighbourhood ranks inform about the order of distances, and are thus an interesting quantity to retain in the framework of the neighbourhood retrieval task. They tend to mitigate the effect of norm concentration in high dimensional data spaces, and intrinsically provide comparable values between the data and embedding space, without the need for scaling. Yet, they tend to be insensitive to changes of points density, or to the presence of gaps in the space, i.e. sudden variations of the step

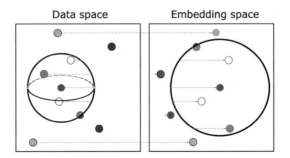

Fig. 3.1 Example of rank distortions with 4-neighbourhoods. The 4-neighbourhood of the black point is materialized in both spaces by the black sphere and circle. Reliable neighbours (white), false neighbours (blue), missed neighbours (red) and non-neighbours (hatched grey) are characterized by their belonging to that neighbourhood in both spaces

in the distances separating a point from successive neighbours. For neighbourhood distortions, assessed through ranks we distinguish:

- *false neighbours* (or neighbourhood intrusions): $\rho_{ij} < r_{ij}$,
- *missed neighbours* (or neighbourhood extrusions): $\rho_{ij} > r_{ij}$.

Mild and hard rank distortions may for instance be distinguished by a quantitative rule, considering a threshold κ. In that case, mild distortions $i \sim j$ are those for which the point j is in the κ-neighbourhood of i in both spaces, while for hard distortions it is in the κ-neighbourhood in only one of the space [105]. If a point j is out of the κ-neighbourhood of i in both spaces, it is called here a non-neighbour and is usually ignored in the assessment of rank distortions. Mildly distorted κ-neighbourhood relations are often considered as reliable, so that only the hard distortions are taken into account (Fig. 3.1).

3.2.2 Link Between Distortions and Mapping Continuity

Distortions of neighbourhood relations may be interpreted in terms of mapping continuity. Intuitively, the continuity of a mapping $\widehat{\Phi} : \mathcal{D} \longrightarrow \mathcal{E}$ between metric spaces ensures that the image of a sufficiently small ball around a point ξ_0 by the mapping is comprised within a ball of given size around the point image $x_0 = \widehat{\Phi}(\xi_0)$. More formally:

Definition 3.1 A mapping $\widehat{\Phi} : \mathcal{D} \longrightarrow \mathcal{E}$, with $(\mathcal{D}, \Delta)$ and $(\mathcal{E}, D)$ metric spaces is said to be continuous in $\xi_0 \in \mathcal{D}$ if for all $\epsilon > 0$, there exists $\omega > 0$ so that for $\xi \in \mathcal{D}$, $\Delta(\xi_i, \xi) < \omega \implies D\left(\widehat{\Phi}(\xi_0), \widehat{\Phi}(\xi)\right) < \epsilon$. This means that for any ball $\mathcal{B}(x_0, \epsilon)$ (centred at $x_i = \widehat{\Phi}(\xi_0)$ and of radius ϵ) in the co-domain, there exists a radius ω such that the image by $\widehat{\Phi}$ of the ball $\mathcal{B}(\xi_0, \omega)$ is included in $\mathcal{B}(x_0, \epsilon)$ (see the illustration in Fig. 3.2).

Fig. 3.2 Continuity of a
function between two metric
spaces (illustration of
Definition 3.1)

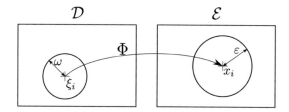

In practice, most DR methods only define a discrete mapping $\Phi : \{\xi_i\} \longrightarrow \{x_i\}$. Thus, the formal concept of continuity may only be applied to an extension $\widehat{\Phi} : \mathcal{M} \longrightarrow \mathcal{E}$ of Φ to the entire data manifold $\mathcal{M}$.

A manifold tear or missed neighbourhood corresponds to a case where a neighbour ξ of a data point ξ_0 (which means a point that would be in "any" ball centred at ξ_0) is not mapped within a ball around the image $x_0 = \widehat{\Phi}(\xi_0)$ of ξ_0. Hence, this type of distortion suggests a breach of continuity of the mapping $\widehat{\Phi}$.

Conversely, a manifold gluing or false neighbourhood implies a breach of continuity for the mapping inverse $\widehat{\Phi}^{-1}$. Indeed, it means that a neighbour x of an embedded point x_0, which is a point that would be in "any" ball centred at x_0, is not mapped within a ball around the image $\xi = \widehat{\Phi}^{-1}(x)$ of x. Note that this relies on the assumption that $\widehat{\Phi}$ admits an inverse.

In that regard,t an ideal mapping, subject to no distortions would be a homeomorphism or bi-continuous function, namely an invertible function that is continuous and whose inverse is continuous. Distortions indicators described in Sect. 3.2.4 assess the breach of continuity for the theoretical mapping $\widehat{\Phi}$ (and its inverse) based on the available information for the discrete mapping Φ, which is its restriction to the sample points $\{\xi_i\}$. For rank-based indicators, this is done by considering the preservation of κ-neighbourhoods, which are balls centred at the points and whose radii are defined by the distance to the κth nearest neighbour of each point.

3.2.3   Reasons of Distortions Ubiquity

In the process of mapping high dimensional data to a low-dimensional space, distortions are present in most cases. First, the best possible map according to the stress function of a given DR technique (i.e. the global optimum of that stress) may not always be reached. Indeed, most of them have non-convex stress functions, so that optimization algorithms may remain stuck at local optima and never converge towards that global optimum (see Sect. 7.1).

Furthermore, some datasets (associated to an hypothetical underlying manifold) may be *non-mappable* in a given embedding space. This means that even the best map may not preserve all relevant information. Assuming that a non-distorted mapping is an homeomorphism (see Sect. 3.2.2), datasets are not mappable in an embedding space, if they live on a manifold that is not homeomorphic to this embedding space. However, in practice, the true underlying manifold associated to a discrete dataset is unknown. Thus, a non-smooth mappable manifold could technically be fitted to any discrete dataset.

If the intrinsic dimensionality of the manifold is strictly higher than that of the embedding space, there exists no homeomorphism between the two spaces. Thus, such a discrepancy is a sufficient condition for the non-mappability of the dataset. This is rather frequent in visual exploration for which the embedding dimensionality is constrained to be very low (at most three for visualization with a single scatter plot). Yet, this condition is not necessary, since some manifolds may not be mapped to embedding space of adequate dimensionality due to topological incompatibility. This is for example the case of a intrinsically two-dimensional sphere, which despite its dimensional compatibility with the two-dimensional plane, is not homeomorphic to it. Note that by puncturing it in a point (introducing a tear), the sphere becomes homeomorphic to a plane (e.g., the Riemman mapping).

3.2.4 Scalar Indicators

Indicators mayt be formulated either as quality indicators or as distortion indicators. Quality indicators increase with the quality of the map and are often normalized to range from zero for the worst mapping to one for an ideal mapping. For distortion indicators, the values increase with the level of distortion, starting from zero indicating the absence of distortion (ideal mapping). They usually compare for a set of pairwise relations, a *reference* distance or rank to its associated *image* distance or rank. For stretching (or missed neighbours) indicators, the reference is the value in the data space and the image is the value in the embedding space (image of the data space by the mapping). For compression (or false neighbours) indicators the reference is taken in the embedding space, and the image in the data space (image of the embedding space by the inverse of the mapping). This section presents several indicators measuring distortions either for global or local map evaluation. For local evaluation, indicators are defined pointwise or pairwise but may be aggregated to a scalar mapwise value (see Sect. 4.2).

3.2.4.1 Distance-Based Indicators

The quality of an embedding may be assessed by comparing the distances Δ_{ij} and D_{ij} between the two spaces. Most indicators from that framework are derived from the normalized "stress" [21], which is both a stress function for DR and used to assess map quality:

$$\mathscr{S} \triangleq \frac{\sum_i \sum_j (\Delta_{ij} - D_{ij})^2}{\sum_i \sum_j \Delta_{ij}^2}.$$

This indicator is however sensitive to scaling of the embedding space distances. Thus, other map evaluation and interpretation criteria tend to independently normalize the distances in both spaces. Yet, a normalization dependent on the embedding distances may differ from one map to another. Hence, it induces additional variability in the evaluation scores.

The projection precision score [160] uses root mean square of distances to the κ nearest neighbours for normalization, in line with the focus of many DR methods on the preservation of small distances:

$$
\mathscr{S}_i^{\text{pps}}(\kappa) \triangleq \sqrt{\sum_{k=1}^{\kappa} \left(\frac{\Delta_{i\tilde{v}_i(k)}}{\sqrt{\sum_{l=1}^{\kappa} \Delta_{i\tilde{v}_i(l)}}} - \frac{D_{i\tilde{n}_i(k)}}{\sqrt{\sum_{l=1}^{\kappa} D_{i\tilde{n}_i(l)}^2}} \right)^2 },
$$

where $\tilde{v}_i$ and $\tilde{n}_i$ are the neighbourhood permutations defined in Sect. 1.1.2. The use of a normalizing value aggregating distances to several points seems to lead to a rather robust indicator. We may note that this normalizing value is also defined pointwise, scaling differently each neighbourhood.

Martins et al. [122] use the following point-wise aggregated error, using absolute values rather than squares:

$$
\mathscr{E}_i \triangleq \sum_{j \neq i} |\mathscr{E}_{ij}| \quad \text{with} \quad \mathscr{E}_{ij} \triangleq \frac{D_{ij}}{D_{\max}} - \frac{\Delta_{ij}}{\Delta_{\max}}. \tag{3.1}
$$

The normalization is performed using the data diameter $\Delta_{\max}$ and map diameter $D_{\max}$ computed as the maximum of all pairwise distances in the two spaces. The use of that single distance is however not very robust since a single outlying point may change the normalization of all other distances. Stretches severity may be assessed with the positive part $\mathscr{E}_{ij}^+$ of the pairwise indicator $\mathscr{E}_{ij}$, and compressions with its negative part $\mathscr{E}_{ij}^-$.

Stress maps [164] use a similar stress, also including a two-parameters weighting strategy:

$$
\mathscr{S}_i^{\text{sm}} \triangleq \sum_{j \neq i} \left(1 - \frac{\Delta_{ij}}{\Delta_{\max}} \right)^{\alpha} \left(1 - \frac{D_{ij}}{D_{\max}} \right)^{a} \left(\frac{\Delta_{ij}}{\Delta_{\max}} - \frac{D_{ij}}{D_{\max}} \right)^2 .
$$

The exponents α and a, both in $[0, +\infty[$, allow to weight the errors based on the type of relation affected. Indeed, increasing α or a respectively leads to giving more importance to small distances in the data or embedding space. Hence, taking $\alpha = 0$ leads to a compression indicator (focusing on distances represented small) and $a = 0$ to a stretch indicator (focusing on originally small distances). In that case, the value of the non-zero parameter is used to tune the neighbourhood scale (relative importance of small and high scales preservation).

3.2.4.2   Rank-Based Indicators

Information Retrieval Approach: Precision and Recall

Precision and recall indicators come from the field of information retrieval. In this framework, the goal is to retrieve, based on a query, the results that are relevant for that query, and only those results. Precision measures the proportion of retrieved results that are effectively relevant. As such, it assesses the ability to retrieve only relevant results and to avoid false positives (retrieved but not relevant), which relates to the specificity of a test. Inversely, recall determines the proportion of relevant results that is effectively retrieved. Hence, it quantifies the ability to retrieve all the relevant results and to avoid false negatives (relevant but not retrieved), which is linked with the sensitivity of a test.

This framework is adapted to map evaluation by considering that the map is a tool for performing the task of (visual) neighbourhood retrieval [189]. Thus, the retrieved results are the nearest neighbours in the embedding space, while the relevant results are the nearest neighbours in the data space, considering hard κ neighbourhoods in both spaces. We may note that the false positives are the false neighbours and the false negatives are the missed neighbours. Hence, the precision $\check{\mathscr{P}}(\kappa)$ and recall $\check{\mathscr{R}}(\kappa)$ of the map (averaged on the neighbourhoods of all points) are given for a neighbourhood size κ by:

$$\check{\mathscr{P}}(\kappa) \triangleq \frac{1}{N} \sum_i \left(1 - \frac{|n_i(\kappa) \setminus v_i(\kappa)|}{|n_i(\kappa)|} \right), \tag{3.2}$$

$$\check{\mathscr{R}}(\kappa) \triangleq \frac{1}{N} \sum_i \left(1 - \frac{|v_i(\kappa) \setminus n_i(\kappa)|}{|v_i(\kappa)|} \right), \tag{3.3}$$

where $| \cdot |$ is the cardinal of a set.

Precision and recall respectively account for the false and missed neighbours. The number of relevant and retrieved neighbours being set at the same value κ, as opposed to the information retrieval case, those indicators are always equal (as illustrated in Fig. 3.3). They are also equivalent to agreement rate [2], also called Local Continuity Meta Criterion (LCMC) [33]. This measure, considers for each point, the size of the intersection between its neighbourhoods in the two spaces. The average value for a given map is:

$$\check{\mathscr{A}} \triangleq \frac{1}{N} \sum_i \frac{|v_i(\kappa) \cap n_i(\kappa)|}{\kappa}.$$

An alternate version of this quality indicator is proposed so that the zero value does not code for the worst possible mapping, but for a random mapping. It is called the corrected agreement rate [2] (equivalent to the adjusted LCMC [33]),

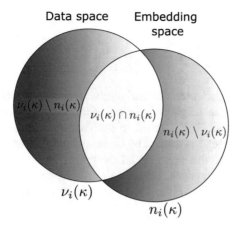

Fig. 3.3 Venn diagram of the κ-neighbourhoods $\nu_i(\kappa)$ and $n_i(\kappa)$ of a point i in the data and embedding spaces respectively. The intersection $\nu_i(\kappa) \cap n_i(\kappa)$ (in white), and the differences $\nu_i(\kappa) \setminus n_i(\kappa)$ (in red) and $n_i(\kappa) \setminus \nu_i(\kappa)$ (in blue) respectively correspond to the reliable, missed and false κ-neighbours. The complement of the union of these sets $(\nu_i(\kappa) \cup n_i(\kappa))^{\complement}$ (white background) corresponds to all the non-neighbours. Considering that the cardinal of $\nu_i(\kappa)$ and $n_i(\kappa)$ is equal to κ, the sets of false and missed neighbours have the same size, which directly depends on the size of the intersection $\nu_i(\kappa) \cap n_i(\kappa)$

and computed as:

$$\check{\mathscr{A}} \triangleq \frac{\check{\mathscr{A}} - \mathbb{E}(\check{\mathscr{A}}_{\text{rand}})}{1 - \mathbb{E}(\check{\mathscr{A}}_{\text{rand}})},$$

where $\mathbb{E}(\check{\mathscr{A}}_{\text{rand}})$ is the expected value of $\check{\mathscr{A}}$ for a random mapping. Assuming that neighbours are randomly drawn without replacement from the entire set of $N - 1$ possible neighbours amongst which only κ are relevant neighbours, the number of reliable neighbours for a given point follows a hyper-geometrical distribution [33]. Thus, this expected value is given analytically by:

$$\mathbb{E}(\check{\mathscr{A}}_{\text{rand}}) = \frac{\kappa}{N - 1}.$$

A similar approach could be used to normalize other rank-based indicators in order to give them a wider range of possible values for mappings found by DR methods, which are almost always better than random mappings, and far better than the theoretical worst mapping.

This also relates with the set difference introduced for local evaluation [123], that corresponds to the Jaccard distance between κ-neighbourhoods $\nu_i(\kappa)$ and $n_i(\kappa)$. This gives the following point-wise distortion measure (using a slightly different normalization compared with agreement rate):

$$\mathscr{J}_i \triangleq 1 - \frac{|v_i(\kappa) \cap n_i(\kappa)|}{|v_i(\kappa) \cup n_i(\kappa)|}.$$

In the above definition, precision and recall are defined by considering κ-nearest neighbours, namely the neighbours in the κ-nearest neighbours graphs in both spaces. They have also been adapted to the Extended Minimum Spanning Tree (EMST) [133]. The EMST is defined by first computing a Minimum Spanning Tree (MST), which is a tree that connects all the points while minimizing the sum of distances associated to its edges. Then, edges are added to connect each point to all its neighbours within a given radius. The radius is defined adaptively for each point based on the properties of the original MST. This leads to the following formulas (defined point-wise):

$$\mathscr{P}_i^{\text{EMST}} \triangleq 1 - \frac{|n_i^{\text{EMST}} \setminus v_i^{\text{EMST}}|}{|n_i^{\text{EMST}}|},$$

$$\mathscr{R}_i^{\text{EMST}} \triangleq 1 - \frac{|v_i^{\text{EMST}} \setminus n_i^{\text{EMST}}|}{|v_i^{\text{EMST}}|},$$

where v_i^{EMST} and n_i^{EMST} are the set of neighbours of point i in the EMST in the data and embedding spaces respectively. This alternative definition has the advantage of not requiring to set a neighbourhood size parameter [133].

Severity Measures: Trustworthiness and Continuity

Trustworthiness $\check{\mathscr{T}}$ and continuity $\check{\mathscr{C}}$ [186] assess at a scale κ the *out-of-neighbourhood rank*, which is the image rank minus the neighbourhood size, for each false and missed neighbours. Aggregated mapwise, this gives the following quality indicators (from zero for worst to one for best):

$$\check{\mathscr{T}}(\kappa) \triangleq 1 - \frac{1}{N} \sum_i \frac{1}{\mathscr{T}_{\max}(\kappa)} \sum_{j \in n_i(\kappa) \setminus v_i(\kappa)} \rho_{ij} - \kappa, \qquad (3.4)$$

$$\check{\mathscr{C}}(\kappa) \triangleq 1 - \frac{1}{N} \sum_i \frac{1}{\mathscr{C}_{\max}(\kappa)} \sum_{j \in v_i(\kappa) \setminus n_i(\kappa)} r_{ij} - \kappa, \qquad (3.5)$$

where $\mathscr{T}_{\max}(\kappa)$ and $\mathscr{C}_{\max}(\kappa)$ are the values of the sums of out-of-neighbourhood ranks (i.e. $\rho_{ij} - \kappa$ or $r_{ij} - \kappa$) on indices j for a theoretical worst mapping. This worst mapping is considered to reverse all ranks, namely, to satisfy $r_{ij} = N - \rho_{ij}$ for all $j \neq i$. The normalizing factors may thereby be obtained by a simple combinatorial analysis:

$$\mathcal{T}_{max}(\kappa) = \mathcal{C}_{max}(\kappa) = \begin{cases} \frac{\kappa(2N-3\kappa-1)}{2}, & \text{if } \kappa < \frac{N}{2}, \\ \frac{(N-\kappa)(N-\kappa-1)}{2}, & \text{if } \kappa \geqslant \frac{N}{2}. \end{cases} \tag{3.6}$$

The definitions of trustworthiness and continuity are close to the ones of precision and recall. Indeed, precision and recall may be obtained replacing the out-of-neighbourhood rank by 1 (and recomputing the associated normalization factor) in Eqs. (3.4) and (3.5). Hence, trustworthiness and continuity account for the *severity* of distortions, namely the level of impact of the distortion on the neighbourhood relation, measured by this out-of-neighbourhood rank error.

Criticality Weighting: Mean Relative Rank Error

Mean Relative Rank Error (**MRRE**) [105] introduces a weighting depending on the importance of the preservation of the considered neighbourhood relation for the overall quality of a mapping. These weights are defined as the inverse of the reference rank. In addition, they differ from trustworthiness and continuity by the use of the rank error (difference between image and reference ranks) instead of the out-of-neighbourhood rank (difference between image rank and neighbourhood size) to penalize distortions. In addition, as opposed to the previous indicators, they account for mild rank distortions. They may be expressed (considering the "unified" version of **MRRE** [105]) as:

$$\check{\mathcal{F}}^{MRRE}(\kappa) \triangleq 1 - \frac{1}{N} \sum_i \frac{1}{\mathcal{F}^{MRRE}_{max}(\kappa)} \sum_{j \in n_i(\kappa) | \rho_{ij} > r_{ij}} \frac{\rho_{ij} - r_{ij}}{r_{ij}}, \tag{3.7}$$

$$\check{\mathcal{M}}^{MRRE}(\kappa) \triangleq 1 - \frac{1}{N} \sum_i \frac{1}{\mathcal{M}^{MRRE}_{max}(\kappa)} \sum_{j \in v_i(\kappa) | r_{ij} > \rho_{ij}} \frac{r_{ij} - \rho_{ij}}{\rho_{ij}}, \tag{3.8}$$

with the normalization factors defined by:

$$\mathcal{F}^{MRRE}_{max}(\kappa) = \mathcal{M}^{MRRE}_{max}(\kappa) = \sum_{k=1}^{\kappa} \frac{\max(0, N - 2k)}{k}. \tag{3.9}$$

The weighting of distortions introduces the idea that all neighbourhood relations are not equivalently critical for the mapping, and that the nearest neighbours are the most important to preserve (in both spaces). Hence, the combination of the severity of distortions (measured by the rank errors) with the associated weights allows to obtain an overall criticality for the mapping of the set of all distortions. Note that in this mindset, non-existing κ-neighbourhood relations may be seen as accounted for with a null weight.

The sequence difference [123] defined for local evaluation also uses rank errors measure of severity and integrates weighting of the different distortions based on the relative importance of their associated neighbourhood relation. This weight is chosen linearly decreasing with the reference rank, smoothly leading to null weights for non-existing κ-neighbourhood relations. It is expressed as:

$$\mathscr{I}_i^{\text{seq}}(\kappa) \triangleq \frac{1}{2} \sum_{j \in n_i(\kappa)} (\kappa - r_{ij})|\rho_{ij} - r_{ij}| + \frac{1}{2} \sum_{j \in v_i(\kappa)} (\kappa - \rho_{ij})|r_{ij} - \rho_{ij}|.$$

This indicator accounts twice for the mild rank distortions, as is the case for the original version of MRRE, which differs from the "unified version" of MRRE (used here) [105]. Indeed, the first sum, mainly assessing false neighbours, also measures mild false missed neighbours (negative values of $\rho_{ij} - r_{ij}$), and the same occurs for mild false neighbours in the second sum. We may note that the sequence difference aggregates by default both types of distortions through an arithmetic mean of sub-indicators.

A "unified version" of the sequence difference could be obtained by defining the false neighbours and missed neighbours quality indicators:

$$\check{\mathscr{F}}^{\text{seq}}(\kappa) \triangleq 1 - \frac{1}{N} \sum_i \frac{1}{\mathscr{F}_{\max}^{\text{seq}}(\kappa)} \sum_{j \in n_i(\kappa)|\rho_{ij} > r_{ij}} (\kappa - r_{ij})(\rho_{ij} - r_{ij}), \qquad (3.10)$$

$$\check{\mathscr{M}}^{\text{seq}}(\kappa) \triangleq 1 - \frac{1}{N} \sum_i \frac{1}{\mathscr{M}_{\max}^{\text{seq}}(\kappa)} \sum_{j \in v_i(\kappa)|r_{ij} > \rho_{ij}} (\kappa - \rho_{ij})(r_{ij} - \rho_{ij}), \qquad (3.11)$$

with

$$\mathscr{F}_{\max}^{\text{seq}}(\kappa) = \mathscr{M}_{\max}^{\text{seq}}(\kappa) = \sum_{k=1}^{\kappa} (\kappa - k) \max(N - 2k, 0).$$

3.2.5   Aggregation

Values of a pair of indicators for several scales are sometimes represented as a bi-dimensional plot showing the evolution of one against the other as presented Fig. 3.4a [185, 189]. In that display, each embedding corresponds to a curve showing the evolution of the two indicators for several scales. This representation may be simplified by aggregating the different indicators allowing to represent the quality directly as a function of the scale (see Fig. 3.4b), or by aggregating the values of a given indicator for several scales, summarizing it by a scalar multiscale value.

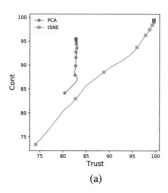
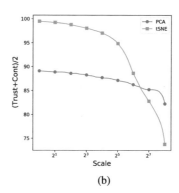

(a)   (b)

Fig. 3.4 Multi-scale representation of pairs of quality indicators for maps of the digits dataset (subsampled) obtained with the PCA and tSNE methods. The left insert plots trustworthiness against continuity (with the scale $\kappa = 1$ identified by a black edge marker), while the right insert shows their arithmetic mean against the scale κ. Scales κ from 1 to $\lceil N/2 \rceil$ are considered. (**a**) Trustworthiness vs continuity. (**b**) $\check{\mathscr{T}}$ and $\check{\mathscr{C}}$ mean vs scale

3.2.5.1 Indicators Aggregation

Indicators derived fromt precision and recall, as the EMST version [133] or the soft precision and recall [189] are sometimes aggregated into the $F-$measure (also called $F1$ score), which is a harmonic mean of the two indicators:

$$F \triangleq 2 \frac{\check{\mathscr{P}} \cdot \check{\mathscr{R}}}{\check{\mathscr{P}} + \check{\mathscr{R}}}.$$

Intuitively, the choice of an harmonic mean may be justified by the fact that the harmonic mean F_i of the pointwise values $\mathscr{P}_i(\kappa)$ and $\mathscr{R}_i(\kappa)$ corresponds to:

$$F_i \triangleq \frac{|v_i(\kappa) \cap n_i(\kappa)|}{\frac{1}{2}(|v(\kappa)| + |n_i(\kappa)|)}.$$

Other rank-based quality indicators defined by pairs $(\check{\mathscr{F}}, \check{\mathscr{M}})$ are sometimes aggregated using the arithmetic mean indicating an average quality, while supplementing this value by another one whose sign indicates the main type of distortion [105]:

$$\frac{\check{\mathscr{F}} + \check{\mathscr{M}}}{2} \quad \text{and} \quad \check{\mathscr{F}} - \check{\mathscr{M}}.$$

3.2.5.2 Scale Aggregation

Choosing the scale parameter κ for a rank indicator is difficult and may appear partial. Some authors choose to present curves showing the evolution of indicators as a function of the scale [48, 107]. Such a multi-scale evaluation may be aggregated by computing the area under the curve [107]. Using a logarithmic scale for the κ axis (which gives more importance to the small scales), this may be computed as:

$$\mathrm{AUC}(\breve{\mathscr{I}}) \triangleq \frac{\sum_{\kappa=1}^{N-1} \breve{\mathscr{I}}(\kappa)/\kappa}{\sum_{\kappa=1}^{N-1} 1/\kappa}.$$

To distinguish between local and global behaviours, the indicators can also be aggregated into two separate values [106]. Considering a threshold κ_0 defined as the best scale for the map, namely the scale maximizing the quality indicator, the preservation of local and global scales are expressed as the arithmetic means of the indicator values for scales respectively below and above κ_0.

3.2.6 Diagrams

To assess the quality of an embedding, the distortion of pairwise relations can also be represented using a diagram. Scalar indicators provide a summary of the information contained in such representations, focusing on certain parts (e.g. small scales or severe distortions). Thus, scalar indicators may be compared more easily between maps, due to the existence of a total order relation in $\mathbb{R}$. Conversely, these diagrams present disaggregated information about distortions, as for local map evaluation (see Sect. 3.2.4). However, in that case, information is not localized on the map and only shows global phenomena in the embedding process.

3.2.6.1 Shepard Diagram

Shepard diagram plots all embedding space distances D_{ij} against the corresponding data space distances Δ_{ij}. For N data points, this leads to N^2 points in the diagram. To avoid overdraw issues, this may also be shown as a histogram of the joint distribution. For an ideal map in the sense of distance preservation, the points of the Shepard diagram are distributed along the diagonal. A less constraining requirement might be that of monotonicity [166]. The level monotonicity may be summarized as a scalar value by using the stress of the Kruskal algorithm [99] or the Spearman rank correlation [60].

Shepard diagram also enables to determine predominant types of distortions for different ranges of distances. Figure 3.5 shows examples of Shepard diagrams for two-dimensional maps of the iris and digits datasets (presented Sect. 1.1.7) obtained

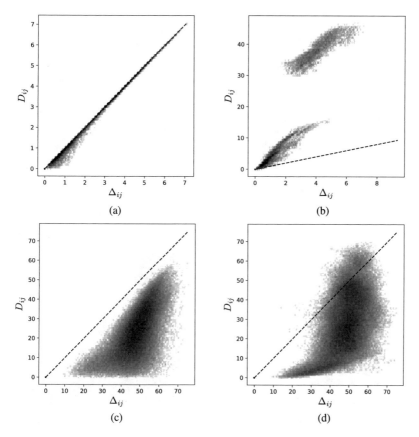

Fig. 3.5 Examples of Shepard diagrams. (**a**) Iris with PCA. (**b**) Iris with tSNE. (**c**) Digits with PCA. (**d**) Digits with tSNE

with PCA and tSNE. Figure 3.5a shows that for the iris dataset the PCA map provides a relatively faithful representation of the distances. For the digits dataset, this shows that PCA (Fig. 3.5c) is compressing the manifold (which is coherent with its nature of linear projection) and that tSNE (Fig. 3.5d) gives a good representation of small distances, but stretches the long distances. Figure 3.5b shows two separate behaviours associated to small and high data distances. These two noisy lines are separated by an offset showing that tSNE tends to stretch high distances (that may be assumed to be between clusters) by a constant for the iris dataset.

3.2.6.2 Co-Ranking Matrix

Thet co-ranking matrix [105] is a $N - 1 \times N - 1$ matrix whose element (ρ, r) gives the number of pairwise neighbourhood relations $i \sim j$ characterized by

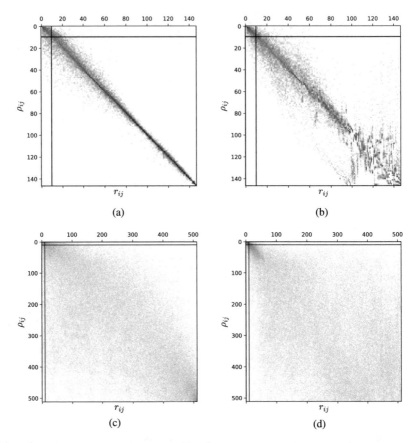

Fig. 3.6 Examples of co-ranking matrices. (**a**) Iris with PCA. (**b**) Iris with tSNE. (**c**) Digits with PCA. (**d**) Digits with tSNE

neighbourhood rank $\rho_{ij} = \rho$ in the data space and $r_{ij} = r$ in the embedding space. It is illustrated Fig. 3.6. Due to the discrete nature of ranks, a natural binning is available to represent this joint distribution. Given a neighbourhood scale κ, the co-ranking matrix may be split into four blocks. The upper left ($i \leqslant \kappa$ and $j \leqslant \kappa$) and lower right ($i > \kappa$ and $j > \kappa$) blocks, namely the diagonal blocks, respectively correspond to the reliable and non-existing κ-neighbourhoods, whereas the lower left ($i > \kappa$ and $j \leqslant \kappa$) and upper right ($i \leqslant \kappa$ and $j > \kappa$) blocks respectively account for the κ-false and missed neighbourhood relations. We may note that the co-ranking matrix is similar to the confusion matrix used to assess the quality of a classifier, replacing classes by rank values.

3.3 Class-Aware Indicators

When class-information is available for the dataset, class-aware indicators allow to measure how the classes are retained by the mapping.

3.3.1 Class Separation and Aggregation

It is sometimes considered that a good dimensionality reduction method should separate the classes well, mapping them to distinct regions of the space. Class-aware indicators based on this paradigm tend to only consider the position of embedded points and the class labels, in order to assess the separation of classes in the embedding space, regardless of the effective data structure.

3.3.1.1 Class Separation

Class separation (also designated as class consistency, homogeneity, purity or uniformity) means that each point is surrounded by neighbours of its class [11, 133, 168]. In that regard, we may define *class separation* as the absence of *class overlap*, where a class overlap corresponds to several classes occupying a common region of space. Strongly overlapping classes may be designated as mixed classes.

A general graph-based framework for measuring class-separation proposes to combine a proximity graph (in the embedding space) and a purity measure, leading to 2002 different indicators with the considered cases [11]. The proximity graph defines a neighbourhood for each point, by the set of points connected to it by an edge, while the purity measure quantifies the amount of intraclass neighbours in this neighbourhood. In that regard, this framework also encompasses pre-existing measures. A simple example of that considering the κ-nearest neighbours graph and the class proportion purity measure is the neighbourhood hit [133]. It computes the average proportion of intraclass neighbours in the κ-neighbourhood of each point and is given by:

$$\breve{\mathcal{H}}_i(\kappa) \triangleq \frac{|n_i(\kappa) \cap \mathcal{S}_i^{\in}|}{|n_i(\kappa)|}, \tag{3.12}$$

with $|\cdot|$ the cardinal of the set. Here, $n_i(\kappa)$ correspond to the neighbours of point i in the κ-nearest neighbours graph, and $\mathcal{S}_i = \{j \neq i \mid L_j = L_i\}$ to the set of points of the same class as i. Yet, the same definition with the Extended Minimum Spanning Tree (EMST) as proximity graph leads to the class separation measure proposed in [133], while replacing the purity measure by the majority vote results in computing the accuracy of a κ-Nearest Neighbours classifier (see Sect. 3.3.1.2).

In a different approach, the density consistency [168] quantifies the level of classes overlap as the entropy of classes distribution in different regions of space. This measure is aggregated map-wise by averaging over all considered regions weighted by the amount of point that they contain.

3.3.1.2   Classification Accuracy in the Map

The class separation approach is often adopted by DR methods that aim at pre-processing the data for performing classification in the embedding space. As such, a common quality indicator is the accuracy of a classifier $\widehat{\Omega}$ in that space, formally given by:

$$\breve{\mathscr{K}}_i \triangleq \frac{|\{i \mid \widehat{L}_i = L_i\}|}{N}, \tag{3.13}$$

with for each point i, the label estimated by the classifier $\widehat{L}_i = \widehat{\Omega}(x_i)$ and the true label L_i. The distortion measure associated to this quality indicator is the classification error.

Assuming convex classes organized around their centroids the distance consistency indicator [168] considers the accuracy of a nearest centroid classifier. Thus, it measures the proportion of points that are closer from the centroid of their class than from the centroid of any other class.

However, in the most common case, $\widehat{\Omega}$ is a leave-one-out k-Nearest Neighbours classifier (k-NN) [79, 145, 174, 184, 207], where k is often set to 1. The accuracy of that classifier accounts for the proportion of points that belong to the majority class of their k-neighbourhood. We may note that the k-NN accuracy measure is subject to a threshold effect due to the winner take all strategy of the k-NN classifier (used to get hard label outputs). With that strategy, a k-neighbourhood containing only one class or many different classes with one having a slight majority lead to an identical prediction by the classifier . The graph-based separation measure with class proportion purity (see Sect. 3.3.1.1) is the soft version of that hard classification.

3.3.1.3   Confusion Matrix

The classification accuracy scores indicate an overall homogeneity of classes in the map. Yet, it does not give details of which class may collide together. Such details may be obtained through a confusion matrix, whose element (i, j) correspond to the proportion of points from class i assigned to class j by the classifier. Thus, diagonal elements present the purity of each class, while the non-diagonal elements assess a level of overlap of the classes. Yet, in the standard case, this is subject to the threshold effect mentioned above.

3.3.1.4   Class Aggregation

Class separation relates to the notion of *class aggregation* which means that classes are regrouped in connected components [133]. Defining a *class split* as the fact that a same class occupy different regions of space not connected together, class aggregation characterizes the absence of class splits. A graph-based measure is also proposed for class aggregation [133], assessing for each point i the proportion of points from its class reachable from i in the proximity graph pruned from its interclass edges. Note that both class separation and aggregation are considered in one given space (data space or map) regardless of the other.

3.3.2 Comparing Scores Between the Two Spaces

Measures of class separation and aggregation assess properties of the map independently from the effective distribution of classes in the data space. In that regard, computing their score in both the data and embedding space might give a sense of whether the classes tend to be under-separated or over-separated by the mapping. The k-NN gain [48] gives this comparison as the difference between the neighbourhood hit in the embedding and data space, in order to estimate the difference of performance of a k-NN classifier in one space or another:

$$\check{\mathscr{H}}'(\kappa) \triangleq \frac{1}{N} \sum_i \left(\frac{|n_i(\kappa) \cap S_i^\in|}{|n_i(\kappa)|} - \frac{|v_i(\kappa) \cap S_i^\in|}{|v_i(\kappa)|} \right),$$

where $v_i(\kappa)$, $n_i(\kappa)$ and S_i respectively denote for a given point i the set of indices of its κ nearest neighbours in the data and embedding spaces and of the points within its class.

Some measure also choose to compare locally these measures between the two spaces, such as the class separation validation [133], which computes for each point the ratio between the minimum and arithmetic mean of the proportion of intraclass neighbours in the EMST in both spaces:

$$F_i \triangleq \frac{\min\left(\frac{|v_i^{\text{EMST}} \cap S_i^\in|}{|v_i^{\text{EMST}}|}, \frac{|n_i^{\text{EMST}} \cap S_i^\in|}{|n_i^{\text{EMST}}|} \right)}{\frac{1}{2}\left(\frac{|v_i^{\text{EMST}} \cap S_i^\in|}{|v_i^{\text{EMST}}|} + \frac{|n_i^{\text{EMST}} \cap S_i^\in|}{|n_i^{\text{EMST}}|} \right)},$$

where v_i^{EMST} and n_i^{EMST} are the equivalent of $v_i(\kappa)$ and $n_i(\kappa)$ replacing the κ-NN graph by the EMST as proximity graph.

We may note that these measure compare between the two spaces the proportions of intraclass neighbours for each point, without considering whether those neighbours are true reliable neighbours.

3.3.3   Class Cohesion and Distinction

Class separation and aggregation indicate in a given space that classes do not overlap and are not split. Yet, in the context of visual exploration, the map is not expected to separate and aggregate the classes but rather to represent faithfully their structure. With that in mind we rather focus on class cohesion and distinction to evaluate class preservation. We respectively define *class cohesion* and *class distinction* for a mapping as the capacity to maintain the existing class aggregation and the existing class separation.

The distortion of the data structure affecting class cohesion and distinction are respectively the intraclass missed neighbours (or tears within classes) and the interclass false neighbours (or gluings between classes). Conversely, interclass missed neighbours and intraclass false neighbours are neutral to the preservation of classes as defined here. Hence, a way of assessing class preservation can be to take advantage of the many indicators designed to assess distortions of the neighbourhood (see Sect. 3.2.4) structure and to restrict them to only account for the distortions of the class structure. In the following we focus on a few rank-based indicators providing a clear separation between a false and a missed neighbours indicator. To quantify class cohesion and distinction, we introduce here class-aware rank based indicators that we derive from common rank-based quality indicators.

For a pair of rank-based quality indicators $(\check{\mathscr{F}}, \check{\mathscr{M}})$ with $\check{\mathscr{F}}$ penalizing false neighbours and $\check{\mathscr{M}}$ penalizing missed neighbours, we define the associated supervised indicators $\check{\mathscr{F}}^{\notin}$ (interclass false neighbours) and $\check{\mathscr{M}}^{\in}$ (intraclass missed neighbours), restricted to the distortions affecting both the neighbourhood and class structure. In terms of associated distortion measures, we get the relations:

$$\mathscr{F} = \mathscr{F}^{\notin} + \mathscr{F}^{\in} \quad \text{and} \quad \mathscr{M} = \mathscr{C}^{\in} + \mathscr{C}^{\notin}, \tag{3.14}$$

where $\mathscr{F}^{\in}$ and $\mathscr{C}^{\notin}$ are the "anti-supervised" indicators restricted to the distortions affecting the neighbourhood structure but not the class structure, and where any quality indicator $\check{\mathscr{I}}$ is given by $\check{\mathscr{I}} = 1 - \mathscr{I}$, with $\mathscr{I}$ its associated distortion measure. This has been introduced in Colange et al. [36] for the specific case of trustworthiness and continuity.

This leads to the following indicators:

- supervised precision and recall (derived from Eqs. (3.2) and (3.3)):

$$\check{\mathscr{P}}^{\notin}(\kappa) \triangleq \frac{1}{N} \sum_i 1 - \frac{\left|(n_i(\kappa) \setminus v_i(\kappa)) \cap S_i^{\notin}\right|}{|n_i(\kappa)|}, \tag{3.15}$$

$$\check{\mathscr{R}}^{\in}(\kappa) \triangleq \frac{1}{N} \sum_i 1 - \frac{\left|(v_i(\kappa) \setminus n_i(\kappa)) \cap S_i^{\in}\right|}{|v_i(\kappa)|}, \tag{3.16}$$

- supervised trustworthiness and continuity (derived from Eqs. (3.4) and (3.5)):

$$\check{\mathscr{T}}^{\notin}(\kappa) \triangleq 1 - \frac{1}{N} \sum_i \sum_{j \in (n_i(\kappa) \setminus v_i(\kappa)) \cap S_i^{\notin}} \frac{\rho_{ij} - \kappa}{\mathscr{T}_{\max}(\kappa)}, \tag{3.17}$$

$$\check{\mathscr{C}}^{\in}(\kappa) \triangleq 1 - \frac{1}{N} \sum_i \sum_{j \in (v_i(\kappa) \setminus n_i(\kappa)) \cap S_i^{\in}} \frac{r_{ij} - \kappa}{\mathscr{C}_{\max}(\kappa)}, \tag{3.18}$$

- supervised MRRE (derived from Eqs. (3.7) and (3.8):

$$\check{\mathscr{F}}^{\mathrm{MRRE}\notin}(\kappa) \triangleq \sum_i \sum_{j \in (n_i(\kappa) | \rho_{ij} > r_{ij}) \cap S_i^{\notin}} \frac{1}{\mathscr{F}_{\max}^{\mathrm{MRRE}}(\kappa)} \frac{\rho_{ij} - r_{ij}}{r_{ij}},$$

$$\check{\mathscr{M}}^{\mathrm{MRRE}\in}(\kappa) \triangleq \sum_i \sum_{j \in (v_i(\kappa) | r_{ij} > \rho_{ij}) \cap S_i^{\in}} \frac{1}{\mathscr{M}_{\max}^{\mathrm{MRRE}}(\kappa)} \frac{r_{ij} - \rho_{ij}}{\rho_{ij}}.$$

That approach could also extend simply to the sequence difference separated in two indicators (see Eqs. (3.10) and (3.11)).

The normalization parameters used are the same as for the unsupervised case (Eqs. (3.6) and (3.9)), allowing to preserve the relation between unsupervised, supervised and anti-supervised indicators given by Eq. (3.14). However, supervised indicators could also be normalized between 0 and 1. To do so, it would be difficult to obtain an analytic expression of the normalizing terms due to the intricacy between class structure and neighbourhood structure. Yet, the value for the theoretical worst mapping may be computed by assigning the following values to the embedding space ranks:

$$r_{ij} = \begin{cases} N - \rho_{ij}, & \text{if } i \neq j, \\ 0, & \text{if } i = j. \end{cases}$$

3.3.4   The Case of One Cluster per Class

The ideal map in terms of class aggregation and separation is constituted of one cluster by class, each cluster being well-separated from all others. In that case, class separation measures such as the neighbourhood hit or κ-NN accuracy return a value of 100% for all scales κ below the minimum number of elements in a class, and tend to decrease for higher scales when the considered neighbourhoods encompass points from other clusters. Similarly an interclass false neighbour indicator $\check{\mathscr{F}}^{\notin}$ also reaches 100% for κ below this threshold, since all embedding space neighbours of any points are intraclass. On the contrary, an intraclass missed neighbours indicator $\check{\mathcal{M}}^{\in}$ is equal to 100% for scales κ higher than the maximum number of elements in a class. In that case, all intraclass points are neighbours in the embedding space.

3.3.5   Intermediate Conclusions

Map evaluation allows to characterize the level of preservation of a certain type of information for a given mapping, and to compare it between different mappings. The considered information, most often considered for a specific scale can be conveyed by distances or neighbourhood ranks, as well as class information. Here we proposed class-aware indicators combining both ranks and class-information in order to assess how well the effective structure of classes is conveyed by a mapping. Though global map evaluation focuses on the assessment of distortions for a complete map, it often relies on the aggregation of more local evaluations. This local evaluation provides detailed information on the reliability of the representation of a given neighbourhood. Hence, it may be used to support map interpretation as detailed in Chap. 4.

Chapter 4
Map Interpretation

An eye for an eye will only make the whole world blind.

Mahatma Gandhi

Map interpretation encompasses several tools enhancing the map with additional information. Some allow to study the link between axes of the data and embedding space, as detailed Sect. 4.1, while other perform local evaluation of distortions deriving concepts of global map evaluation, as described in Sects. 4.2 and 4.3.

4.1 Axes Recovery

A reliable map of a high dimensional dataset allows to determine the level of similarity or dissimilarity between data points. Yet, for multidimensional data, it may also be interesting to explain the observed dissimilarity between pairs of data points by studying their differences in terms of data features. This may be seen as considering the relation between the value of those data features and the position of embedded points in the map. Finally, that study of the link between data features and synthetic features (i.e. map coordinates), may lead to give a meaning to the synthetic features.

4.1.1 Linear Case: Biplots

In the case of a linear projection technique, such as Principal Component Analysis (**PCA**) [94, 141], the synthetic features are linear combinations of the data features. This is exploited by the biplot view [72, 94] which represents on the map both the embedded points and vectors associated to the data features. Figure 4.1 shows a biplot view for the normalized iris dataset. The norm of a data feature vector qualitatively indicates the amount of variance well-represented by the embedding for that feature. For normalized data, each data feature has unit variance, so that

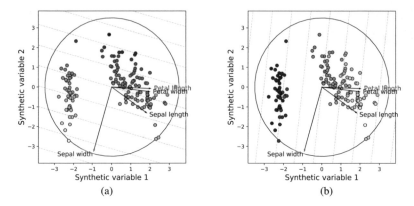

Fig. 4.1 Biplot view for the normalized iris dataset with successive focus on the repartition of two features (sepal width and petal width). The circle indicates the unit norm for the data feature vectors. The dashed lines would be the lines of iso-values for a perfectly represented feature. The the feature values are shown in colour with the viridis colour ramp. (**a**) Sepal width value. (**b**) Petal width value

the norm is upper bounded by one. Hence, if the norm of a data feature vector is close to one, it means that differences between data points for that feature are likely to induce differences of position in the embedding space for these points. As such, Fig. 4.1 shows that the sepal width is very well-represented by PCA and is better-represented than other data features. Furthermore, the angle between a data feature vector and the direction of a synthetic feature characterizes the correlation between these features. Thus, the value of a data feature tends to increase along the direction of its vector, as may be observed for sepal width in Fig. 4.1a. Yet, for data features that are not well-represented, additional variations may occur in direction not accounted for by the embedding. Thereby, for petal width, variations independent from the direction of the feature vector may be observed in Fig. 4.1b. Finally, for a data feature vector with nearly null norm, there would be almost no correspondence between the value of the data feature and the position of points in the map.

4.1.2 Non-Linear Case

For non-linear mappings, it is not possible to determine a global direction in the map along which a data feature tends to increase linearly. Instead, the variation of data features with points position could be characterized locally at each point of the map. The DimReader method [65] determines for each data point how a little perturbation for a given data feature impacts the values of the synthetic features, that is the position of its associated map point. In order to do so, it uses automatic differentiation to compute derivatives of the embedding points coordinates with respect to the position of each data point, given the non-linear DR technique which produced the map. This provides a vector field indicating the direction of displacement of embedded points with such perturbations. Then, the result is displayed on the map using the contour lines orthogonal to the field at each point. Those contour lines provide the iso-values of the data feature as in Fig. 4.1a.

Similarly, for a given data feature, the probing projection tool [170] associates to each embedded point the value of this data feature for its high-dimensional counterpart (as shown by the marker colours in Fig. 4.1). Then, these values are interpolated on the map background, providing a heatmap of that feature. Thereby, it is possible to examine how that value varies across the map and to guess its impact on the proximity or remoteness of embedded points.

Glyphboard [95] allows to consider all original dimensions simultaneously by representing them on the point markers using glyphs, that is symbols with branches whose length indicates the magnitude of the variable. Comparing two glyphs, it is easy to determine which data features explain the difference between two points, but it is harder to get an idea of how a given data feature varies across the map. For image data, for which each data feature is the intensity of a pixel, a similar effect can be reached replacing the glyphs by the images.

4.2 Local Evaluation

Global evaluation indicators detailed Chap. 3 can be used by analysts to select the most appropriate embedding of a dataset in the range of many obtained with different dimensionality reduction techniques and different values of those techniques hyper-parameters. A well-chosen indicator allows to consider a map with the minimal amount of distortions) (assuming a certain definition of distortions). Yet, that selection does not guarantee that the considered map is exempt of any distortions. Indeed, most algorithms are not certain to reach a global optimum and some datasets are even non-mappable (see Sect. 3.2.3).

Hence, while interpreting this map, it is necessary to locate the remaining distortions using local evaluation, to discard misleading visual information and to reduce mapping induced biases. This local evaluation is displayed onto the map providing interpretative support (also referred to as layout enrichment [139] or projection probing [170]). A simple analogy to justify the necessity of interpretative support is that of a two-dimensional map of the Earth surface. Indeed, to read such a map, the viewer must be aware of the structure distortions, which comprise a tear along a meridian passing through the pacific ocean for the Earth map commonly used in Europe.

This local evaluation may be related to global evaluation (see Chap. 3) by noticing that most global indicators rely on aggregation of local assessment of distortions. Thus, we may distinguish three levels of aggregation of distortion measures:

- The mapwise level (see Chap. 3) providing a scalar value $\mathscr{I}$ measuring the total amount of distortions in the map.
- The pointwise level (see Sect. 4.2.1) assessing with an indicator $\mathscr{I}_i$ defined for each point i how much its relations to other points are distorted.
- The pairwise level (see Sects. 4.2.2 and 4.2.3), considering the amount of distortion $\mathscr{I}_{ij}$ for each individual neighbourhood relation i, j, which is the elementary level often aggregated by the previous two.

4.2.1   Point-Wise Aggregation

Local evaluation of distortions defined point-wise computes for each point i a value $\mathscr{I}_i$ characterizing the distortion of its neighbourhood relations by the mapping, based on a given indicator $\mathscr{I}$. Values are then displayed on the map at the position of their associated points, either through point markers or by colouring the background.

4.2.1.1   Point Markers

The severity of distortion can be encoded using different graphic variables associated to the points markers. It can be the colour [129, 133] or the size of a halo placed behind it [170]. In the latter case, the halo may be gray or white to show the sign of the indicator, hence giving the main tendency (compression or stretching). With such neutral colours, the colour channel remains available to encode other elements such as class information. However, the point markers are often small compared to the area in the map. Hence, other methods choose to display the distortions on the map background.

4.2.1.2   Background Colouring

A simple way to associate a region of the background to a point is to compute the Voronoï diagram and to colour the Voronoï cell associated to each point depending on the distortion severity $\mathscr{I}_i$ [10, 112]. The value $\mathscr{I}_i$ may also be interpolated on each pixel of the background presenting them as a heatmap and/or heightmap [160, 164], as illustrated Fig. 4.2. Yet, in very dense areas, the colour change may have a too high spatial frequency, making it difficult to read. Moreover, border points may influence very large regions of the map, giving the undue impression that their associated value is representative of many points. To limit these issues, values may be smoothed and colouring may be restricted to an area within a certain distance of points, using an opacity linearly decreasing with the distance to the nearest embedded point [122, 123].

Pointwise aggregation gives a general idea of the preservation of each point's neighbourhood, but it lacks *directionality*. Indeed, the view of Fig. 4.2 enables

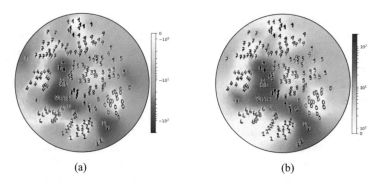

(a)   (b)

Fig. 4.2 Colouring of the background with pointwise aggregation of trustworthiness and continuity signed penalizations as detailed in MING uniform framework (Sect. 4.3.1), for a map of the digits dataset (200 samples) embedded with DD-HDS [114]. Values are interpolated with Sibson interpolation method [167] following the concept of CheckViz [112]. (**a**) Trustworthiness penalization. (**b**) Continuity penalization

to determine whether at least one neighbourhood relation is distorted. However, it is not possible to determine which neighbourhood relations are affected by those distortions. In addition, the same level of pointwise distortion could come from many mild distortions or a few severe distortions. In the latter case, part of the neighbours might be reliably represented, and should be identified. This directionality issue may be solved by considering individually the neighbourhood relations at the pairwise level.

4.2.2 One to Many Relations with Focus Point

Distortions of pairwise relations are often presented for a specific focus point i, allowing to characterize the preservation of all its neighbourhood relations $i \sim j$ without aggregation. Such display of one to many relations usually relies on interactive selection of a focus point to switch between N possible views associated to each of the map points. To avoid considering every single point, a common strategy is to restrict the considered focus points to those with the most distorted neighbourhoods. Those may be identified using an associated point-wise aggregated display of distortions [122, 170].

In the focus point strategy, pairwise relations may be displayed using the background colouring approaches (see Sect. 4.2.1.2) employed for point-wise aggregated indicators [10, 122]. This operation only requires to identify clearly the focus point i and to display for each point $j \neq i$ the pairwise value $\mathscr{I}_{ij}$ instead of the point-wise value $\mathscr{I}_j$. The focus point may be highlighted using a bigger marker [86], a reticle [122] or a marker coloured differently [10, 170].

In addition, for a given focus point i, it is always possible to perfectly represent all distances from i to any other point. Thus, a distance correction approach [170] may present visually the data space distances Δ_{ij} between i and other points $j \neq i$, while also comparing them to their associated distances D_{ij} in the map. To do so, each map point x_j is moved along the line connecting x_i and x_j, and positioned at a distance Δ_{ij} from x_i. The segment between the original map position and the corrected position is then coloured in gray or white depending on whether the point had to be moved away or closer from the focus point, thus distinguishing compression or stretching. This colour code is coherent with the associated point-wise aggregated representation (halo representation described Sect. 4.2.1).

Similarly, Proxilens [86], which is based on a magic lens approach, displaces false neighbours away from the focus point (with an animation), placing them on the circle delimiting the lens. The magic lens is defined as a focus area in which irrelevant information is filtered out. Conversely, the true distance to the reliable and missed neighbours is shown by background colour interpolation, while the background around the false and non-neighbours is coloured in black.

Finally, the pairwise relations may be represented by colour or transparency of edges linking the focus point i to other points $j \neq i$. This approach, used by the one to many missing neighbour finder [122], is a restricted case of the many to many missing neighbours finder detailed in Sect. 4.2.3.2.

4.2.3 Many to Many Relations

Pointwise aggregation provides information on the preservation of the neighbourhood of any given point. Yet, it tends to hide specific information about which pairwise relations are distorted and which are represented faithfully. Conversely, focus-point strategies considers those pairwise relations, but they require to consider many different views to get an idea of the distortions of the global structure. Many-to-many relations views attempt to present simultaneously the level of distortion $\mathscr{I}_{ij}$ for all relevant neighbourhood relations $i \sim j$.

4.2.3.1 Matrix View

Preserving the distances may be seen as minimizing the difference between the distance matrices in the data and embedding spaces. Thus, the **Compadre** tool [44] choose to provide local evaluation as a matrix view accompanying the map. Each entry (i, j) of that matrix represents the level of distortion of the associated neighbourhood relation, with indices of points set using a re-ordering strategy. Hence, all N^2 neighbourhood relations may be displayed simultaneously. The level of distortion is assessed by the distance difference, and mapped to a colour using a diverging red blue colour ramp. Hence, the matrix is constituted of white cells for well represented distances, blue cells for compressed distances and red cells for stretched distances. As opposed to other interpretative support methods, this one does not present distortions directly on the map. In order to link information given by the matrix and position of points in the map, the rows and columns of the matrix may be associated to clusters also shown on the map by point markers. The reordering by clusters may lead to a block structure of the matrix, and help to see the global distortion tendency for the relations between two given clusters.

4.2.3.2 Graph Display

Pairwise relations may also be represented as edges of a graph superimposed onto the map, encoding the distortion measures through edges properties. Martins et al. [122] use two distinct graphs for identifying manifold compression and stretching, with a distance-based indicator. The graph vertices are positioned at the embedding points coordinates, while the distortion of the relation $i \sim j$ are encoded by colour and transparency of the associated edge (i, j). The two types of distortions are assessed with the positive part $\mathscr{E}_{ij}^{+}$ and negative part $\mathscr{E}_{ij}^{-}$ of the indicator $\mathscr{E}_{ij}$ of Eq. (3.1).

To avoid presenting too many relations, the compressions are shown only for edges of the Delaunay graph (whose edges connect points with adjacent Voronoï cells), computed for the map points. This graph allows to consider for each point the pairwise relations with a subset of its map neighbours without any edge crossings. Conversely, stretching is shown only for a user-set proportion of the most stretched pairwise relations of the map. That second graph is strongly affected by edges crossings and overlaps, since stretched point pairs are far apart in the map. To reduce the visual clutter of those edges, a bundling is performed using the Kernel Density Estimation-based Edge Bundling (**KDEEB**) algorithm [91].

4.3 Map Interpretation Using Neighbourhood Graphs

This section describes a new many-to-many method for interpretative support called Map Interpretation using Neighbourhood Graphs (**MING**), that we previously introduced in Colange et al. [35]. This method provides a local and quantified evaluation of distortions. For that purpose it relies on:

- pairwise distortion indicators extracted from global rank-based quality indicators,
- neighbourhood graphs presenting only the relations that are relevant for the interpretation.

Conversely to the many-to-many approach proposed by Martins et al. [122], **MING** focuses on key information for the neighbourhood retrieval task rather than most distorted information. Indeed, at a given scale κ, it shows the reliable proximities (white edges), unreliable proximities (blue edges) and unreliable remoteness (red edges) perceived by an analyst between pairs of points.

4.3.1   Uniform Formulation of Rank-Based Indicators

MING considers pairwise distortion measures used to define pairs of rank-based map quality indicators $(\check{\mathscr{F}}, \check{\mathscr{M}})$, penalizing false and missed neighbours respectively. As such, it can help to interpret the scalar values of those indicators, obtained by aggregation of local measures. Several of those indicators may be expressed in a uniform framework, where the mapwise measure of quality $\check{\mathscr{I}}(\kappa)$ of indicator $\check{\mathscr{I}}$ (either $\check{\mathscr{F}}$ or $\check{\mathscr{M}}$) is given for a neighbourhood size κ by:

$$\check{\mathscr{I}}(\kappa) = 1 - \frac{1}{N} \sum_i |\mathscr{I}_i(\kappa)|.$$

The point-wise distortion measures $\mathscr{F}_i(\kappa)$ and $\mathscr{M}_i(\kappa)$ aggregated mapwise are expressed as:

$$\mathscr{F}_i(\kappa) = \frac{1}{\mathscr{F}_{\max}(\kappa)} \sum_{j \in n_i(\kappa)} w_{ij}^{\mathscr{F}}(\kappa) \, \mathscr{F}_{ij}(\kappa), \tag{4.1}$$

and

$$\mathscr{M}_i(\kappa) = \frac{1}{\mathscr{M}_{\max}(\kappa)} \sum_{j \in v_i(\kappa)} w_{ij}^{\mathscr{M}}(\kappa) \, \mathscr{M}_{ij}(\kappa), \tag{4.2}$$

where the measures of pairwise distortions are the product of the distortion severity $\mathscr{F}_{ij}(\kappa)$ and $\mathscr{M}_{ij}(\kappa)$ and criticality weights $w_{ij}^{\mathscr{F}}(\kappa)$ and $w_{ij}^{\mathscr{M}}(\kappa)$. Normalization factors $\mathscr{F}_{\max}(\kappa)$ and $\mathscr{M}_{\max}(\kappa)$ are the maxima of the sums on j, obtained for a theoretical worst mapping leading to reversed ranks.

This uniform framework encompasses precision and recall [189], trustworthiness and continuity [186] and Mean Relative Rank Error (MRRE) [105]. It may also be extended to the sequence difference distortion indicator [123], with the slight adaptations proposed in Eqs. (3.10) and (3.11). Table 4.1 summarizes the values of distortion measures and normalization factors for those indicators. Note that in this formulation pairwise and pointwise distortion measures are signed to distinguish

Table 4.1 Pairwise distortion measures and normalization factors used in Eqs. (4.1) and (4.2) to define rank-based map quality indicators

$(\mathscr{F}, \mathscr{M})$ pair	$\mathscr{F}_{ij}(\kappa)$ for $j \in n_i(\kappa)$	$w_{ij}^{\mathscr{F}}(\kappa)$	$\mathscr{M}_{ij}(\kappa)$ for $j \in v_i(\kappa)$	$w_{ij}^{\mathscr{M}}(\kappa)$	$\mathscr{F}_{\max}(\kappa) = \mathscr{M}_{\max}(\kappa)$ for $\kappa < \frac{N}{2}$
$(\mathscr{P}, \mathscr{R})$	$-1_{j \in v_i(\kappa)}$	1	$1_{j \in n_i(\kappa)}$	1	κ
$(\mathscr{T}, \mathscr{C})$	$-(\rho_{ij} - \kappa)^+$	1	$(r_{ij} - \kappa)^+$	1	$\frac{\kappa(2N - 3\kappa - 1)}{2}$
$(\mathscr{F}^{\mathrm{MRRE}}, \mathscr{M}^{\mathrm{MRRE}})$	$-(\rho_{ij} - r_{ij})^+$	$\frac{1}{r_{ij}}$	$(r_{ij} - \rho_{ij})^+$	$\frac{1}{\rho_{ij}}$	$\sum_{k=1}^{\kappa} \frac{(N - 2k)}{k}$
$(\mathscr{F}^{\mathrm{seq}}, \mathscr{M}^{\mathrm{seq}})$	$-(\rho_{ij} - r_{ij})^+$	$\kappa - r_{ij}$	$(r_{ij} - \rho_{ij})^+$	$\kappa - \rho_{ij}$	$\sum_{k=1}^{\kappa} (\kappa - k)(N - 2k)$

false neighbours indicators (negative) and missed neighbours indicators (positive). In the next Section, a graph representation of these pairwise elements is proposed.

4.3.2 MING Graphs

Adopting the neighbourhood retrieval perspective [189], we consider that a data analyst uses a map to infer the data space neighbours (relevant neighbours) of each point relying on its map neighbours (retrieved neighbours). The combination of many local inferences may lead to the identification of more global data structures such as clusters or continua. To assess the reliability of those inferences, MING displays two graphs, whose edges represent neighbourhood relations between specific pairs of points, and are coloured depending on the level of distortion of those relations. The two graphs are:

- the retrieval graph, linking each point to its neighbours retrieved by an analyst in the map (either reliable or false neighbours),
- the relevance graph, linking each point to its relevant neighbours effectively found in the data space (either reliable or missed neighbours).

The non-neighbours are not taken into account by those graphs. Formally, the retrieval and relevance graphs are defined as:

$$G_{\mathscr{F}}^{\text{retr}}(\kappa) \triangleq \left((V, E^{\text{retr}}(\kappa), W_{\mathscr{F}}^{\text{retr}}(\kappa)\right) \quad \text{and} \quad G_{\mathscr{M}}^{\text{rel}}(\kappa) \triangleq \left(V, E^{\text{rel}}(\kappa), W_{\mathscr{M}}^{\text{rel}}(\kappa)\right),$$

where V is the set of vertices, E^{retr} and E^{rel} the sets of edges, while $W_{\mathscr{F}}^{\text{retr}}(\kappa)$ and $W_{\mathscr{M}}^{\text{rel}}(\kappa)$ are the sets of weights. The vertices in V correspond to the points and are laid out using the embedded points coordinates.

The set $E^{\text{retr}}(\kappa) \triangleq \{(i, j) \in V \times V \mid j \in n_i(\kappa)\}$ is the set of edges of the directed κ-nearest neighbours graph in the embedding space, and $E^{\text{rel}}(\kappa) \triangleq \{(i, j) \in V \times V \mid j \in \nu_i(\kappa)\}$ is its counterpart in the data space. The edges are represented as polylines between the points, obtained with edge bundling (as detailed in Sect. 4.3.5.2). The directed edge (i, j) is represented as the half of the polyline closer to i. If its symmetric edge (j, i) does not exist in the edges of the graph, the second half of the polyline is represented with the same characteristics as edge (i, j), but dashed. Edges only existing in one direction in the graph may characterize outliers.

Edge properties are defined by the sets of weights $W_{\mathscr{F}}^{\text{retr}}(\kappa)$ (respectively $W_{\mathscr{M}}^{\text{rel}}(\kappa)$), which contain for each edge (i, j) the couple $(\mathscr{F}_{ij}, w_{ij}^{\mathscr{F}})$ (respectively $(\mathscr{M}_{ij}, w_{ij}^{\mathscr{M}})$). The distortion severity is encoded with colours, using YlGnBu and YlOrRd colour ramps [27] for the false and missed neighbours, so that reliable edges appear white, while false neighbours and missed neighbours edges are respectively blue and red. Conversely, the criticality of the relationship is mapped to the line width.

Other minor technical choices allow to highlight the information. A gray background is added to make white edges more visible. The colour scales are logarithmic between 1 and N to increase the colour gradient (with a linear part between 0 and 1). Edges with more severe distortions are drawn on top. Note that the graphs are rendered using the Python library Matplotlib [90].

4.3.3   MING Analysis for a Toy Dataset

We first apply MING to the two rings toy dataset [54], constituted of two circle-like structures intertwined in a three-dimensional space and represented Fig. 4.3a. This dataset is embedded in a two-dimensional space with tSNE [183], leading to the map shown Fig. 4.3b. White edges on both the retrieval and relevance graph show the reliable proximity, allowing to identify two map continua. The blue edges of the retrieval graph (Fig. 4.3c) indicate misleading proximities, showing that the rings are not connected together in the data space despite their closeness in the map. Finally, the red edges show pairs of points unduly distanced in the map, allowing to infer that the continua are in fact closed cycles, though their extremities are torn in the map.

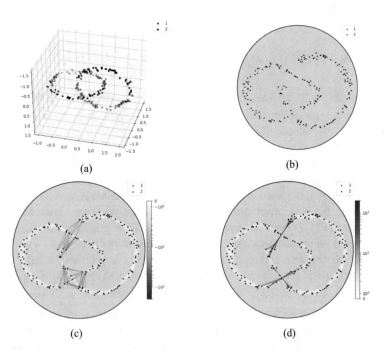

Fig. 4.3 MING for the two rings dataset embedded with tSNE [183], with $\kappa = 15$ and trustworthiness and continuity distortion measure. (**a**) 3D dataset. (**b**) 2D map. (**c**) Retrieval graph. (**d**) Relevance graph

Thereby, MING enables the detection of two disconnected cyclic one-dimensional structures in the data. In addition, the reliable edges highlight the structure that should be visually identified in the map by the analyst (Fig. 4.3b) .

4.3.4 Impact of MING Parameters

We then study the effect of MING parameters for a map of the digits dataset (with 200 random samples) obtained using DD-HDS[114]. The results of MING for that map are shown in Figs. 4.4 and 4.5.

4.3.4.1 Digits Interpretation

Focusing on the retrieval graph enables to distinguish reliable map clusters strongly connected by white edges and linked to their surroundings by blue edges. We may, for instance, identify clockwise starting from the bottom right of the map, map clusters of 9s, 2s, 6s, 4s, 7s, 1s, 4/9s, while in the remaining central area map clusters of 3s, 5s, 8s and 9s may be determined. MING also allows to identify reliably represented interfaces between those map clusters, such as between the 1s and 4/9s map clusters at the top of the map. One may also locate map outliers, such as the lone pair of 2s above the cluster of 6s (and magnified in Fig. 4.5). Those are connected together by a white edge, and to all their map neighbours by blue edges. Considering the relevance graph, one may see torn clusters, as the cluster of 5s, splitted in two subclusters in the map with some 3s in between. This also shows the link between the outlying pair of 2s and the main map cluster of 2s.

Compared with point-wise aggregated view, MING provides directional information. This may be observed by considering Fig. 4.2, which is the aggregated equivalent of the $\kappa = 10$ case in Fig. 4.4. In Fig. 4.2 (focusing on the false neighbour indicator), interface points at the border of map clusters, such as the 2s close to the map cluster of 6s, appear with a high level of distortion, which implies that they are not well-represented. Yet, considering MING view, we may determine that those points are both linked to reliable neighbours (the other 2s) and to unreliable neighbours (false neighbours in the cluster of 6s). For the missed neighbours indicator, this is even more significant, since edges of the relevance point towards the missed neighbours of the considered point.

4.3.4.2 Impact of the Scale

The number of neighbours κ defines a scale of interest for MING graphs. It mainly impacts the number of edges displayed on the map, but also, the measure of distortion severity in the case of $(\mathscr{P}, \mathscr{R})$ and $(\mathscr{T}, \mathscr{C})$ indicators. In that case, more edges appear reliable when κ increases. As for global map evaluation, the choice

Retrieval graph   Relevance graph

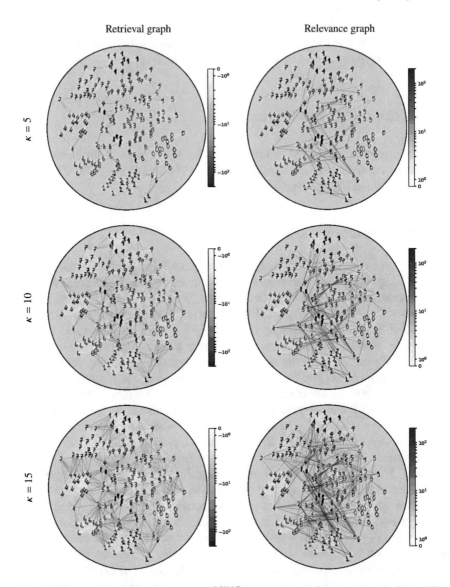

Fig. 4.4 Impact of the scale parameter κ of MING for the trustworthiness and continuity weights and the digits embedded with DD-HDS

of the scale parameter remains subjective. The evolution of MING graphs with the parameter κ is presented Fig. 4.4. Considering the case of $\kappa = 5$, we see that very few neighbourhood relations are characterized in the map. This allows to focus on very important relations, but reduces the ability to assess the reliability of certain proximities, for instance between some map clusters. On the other hand, $\kappa = 15$ provides information about more neighbourhood relations, but leads to a cluttered graph, especially in the case of the relevance graph.

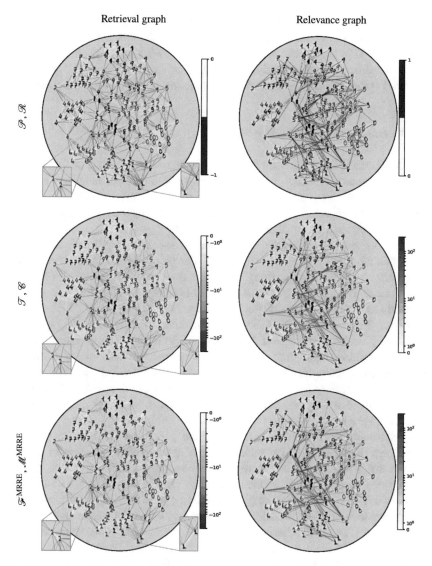

Fig. 4.5 Impact of the type of distortion measure for $\kappa = 10$

The first conclusion is that the appropriate value of κ (which could be set here to $\kappa = 10$) is subject to a trade-off: it must be sufficiently high to provide the necessary information (as opposed to the case of $\kappa = 5$), but also sufficiently small to remain readable. Yet, one may also decide to consider successively several scales to get a basic idea of the core relations at a small scale, and then to confirm the tendency with more edges. This may, for instance, be attained by interactive selection of the value κ.

We may also note that compared with pointwise aggregation, the display of a graph gives an intuitive idea of the extent of the neighbourhood for the considered value of κ. This may help to interpret the mapwise value of the associated quality indicator, since the effective neighbourhood size is not necessarily obvious, particularly when defined by a number of neighbours rather than a fixed radius.

4.3.4.3   Impact of the Distortion Measure

The considered indicator defines how distortions are measured. It is decomposed into distortion severity, which assesses the gravity of the distortion for one pairwise relation, and criticality weight, which accounts for the importance of preserving this relation for the mapping. These are respectively encoded by the colour and linewidth of edges, which leads to the distinct views presented Fig. 4.5.

Distortion severity mostly depends on the image rank (ρ_{ij} for $\mathscr{F}$ and r_{ij} for $\mathscr{M}$), which may vary between 0 and $N-1$. For precision and recall, the severity is a binary measure, distinguishing between values of the image rank higher or lower than κ. Thus, mild distortions slightly above this threshold are represented identically to the most severe distortions, and very differently than the absence of distortions. Thus, some edges inside clusters, though mildly distorted, appear as significantly distorted, and hinder the process of identifying reliable map clusters.

Trustworthiness and continuity introduce more nuances, with a more continuous severity measure, computed as the positive part of the out-of-neighbourhood rank. This allows to distinguish mild and hard distortions. By decreasing the contrast between mildly distorted and reliable distortions, it also facilitates the discovery of reliable map clusters.

MRRE distortion measure uses the positive part of the rank error, which is independent of κ. However, for small values of κ the severity measure of MRRE does not differ significantly from that of trustworthiness and continuity. Indeed, for these two severity measures, the image rank varying from 0 to $N-1$ is either compared to the reference rank ranging from 0 to κ or to the neighbourhood size κ. Yet, in the case of MRRE the criticality weights vary with reference ranks (r_{ij} for $\mathscr{F}$ and ρ_{ij} for $\mathscr{M}$). Those weights being encoded by edges width, edges between close map neighbours in the retrieval graph and close data neighbours in the relevance graph appear thicker than the others in the case of MRRE distortion measure. MING for the sequence difference indicator would lead to a similar figure, but with widths decreasing linearly instead of inversely.

4.3.5   Visual Clutter

One of the key issues of displaying pairwise relations is the readability and scalability of the process. When the number of points N, the number of neighbours κ or the level of distortion increases, MING graph becomes more difficult to read due

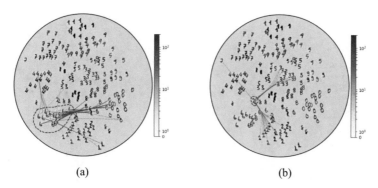

(a)   (b)

Fig. 4.6 Interactive edge filtering for MING relevance graph with $\kappa = 20$ and trustworthiness and continuity distortion measure. The lasso selection is shown by the dashed blue line. (**a**) Selection of the cluster of 6s. (**b**) Selection of the outlying 2s

to visual clutter of the representation. Indeed, the many edge crossings and overlaps make it difficult to follow individual edges. This clutter is especially significant for the unreliable edges of the relevance graph, since they connect points that are distant in the map. Visual clutter may be alleviated relying both on interactive edge selection and edge bundling techniques.

4.3.5.1   Interactive Edge Filtering

When exploring the map, an analyst may successively focus on different groups of points. To ease the study of individual edges starting from a given group, interactive edge filtering enables to momentarily hide all the other edges, thus strongly reducing the impact of visual clutter. As such, an analyst may navigate the graph, alternating between views of the full graph, providing information on the global structure of data, and views of subsets of edges to get more local details.

The selection of points may be performed interactively using lasso selection. This selection may also be performed on the retrieval graph to detect reliable clusters and then investigate relations with their missed neighbours. Figure 4.6 shows the use of that technique to focus on the map cluster of 6s and the outlying pair of 2s. This latter case, using high values of κ, would be very similar to the one-to-many interpretative support methods. The analysis of the cluster of 6s show similarities with the 0s, that are not rendered by the map. As for the the pair of 2s, it is mostly linked to the main cluster of 2s, but also to the cluster of 3s, which might help explain that the DR algorithm placed them in between those two groups.

4.3.5.2   Edge Bundling

A possible approach to reduce visual clutter is to perform edge bundling on the missed neighbours edges of the relevance graph (i.e. edges with maximum clutter

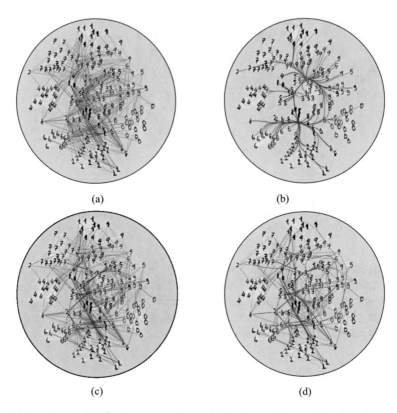

(a)   (b)

(c)   (d)

Fig. 4.7 Bundling of MING relevance graph penalized edges, with $\kappa = 15$ and trustworthiness and continuity distortion measure. (**a**) No bundling. (**b**) Density-based bundling. (**c**) Force-directed bundling. (**d**) Bundling by community

impact). Edge bundling enables to locally aggregate the edges to reduce the amount of space that they take in the visualization of a graph. Different solutions of bundling are presented Fig. 4.7 for the unbundled graph of Fig. 4.7a. In that approach, edges are not represented as straight lines, but as polylines defined by several control points. The control points at the extremities of each edge (i.e. its vertices) are fixed.

Density-Based Bundling

A well-known and computationally efficient method for edge bundling is the Kernel Density Estimation-based Edge Bundling (**KDEEB**) [91]. Its approach is close to the mean-shift clustering algorithm [38] applied to the control points. Each iteration consists in the following steps. First, the probability density function of the distribution of control points is estimated using kernel density estimation (with a given kernel and bandwidth). Then, the control points are advected towards the modes of that distribution (i.e. along the direction of the probability density function gradient). Finally, each edge is smoothed and resampled to reduce the impact of potential numerical artefacts. Since **KDEEB** focuses on each edge independently (using an image splatting technique to store the density estimate), its complexity is $O(|E^{\text{retr}}(\kappa)|)$.

With that algorithm, the edges are locally grouped with other close edges, regardless of their directions, as shown in Fig. 4.7b. In that figure, the missed neighbour edges of **MING** relevance graph are bundled with orthogonal edges, so that it becomes nearly impossible to follow an individual edge. Thereby, **KDEEB** tends to hinder the directionality of information, which is essential for **MING**.

Force-Directed Bundling

Conversely, force-directed approaches account for edges compatibility, to avoid the grouping of incoherent edges. The Force-Directed Edge Bundling (**FDEB**) [89] uses a mechanical simulation in which control points of different edges are attracted towards each other by electromagnetic-like forces (decreasing when the distance increases), while each edge is maintained coherent by the application of spring-like forces (increasing with spring elongation) between successive control points. The magnitude of the electromagnetic force exerted between two control points is weighted by the compatibility of their edges. This compatibility depends on their respective directions, lengths and positions, and prevents the bundling of dissimilar edges (e.g., orthogonal edges). This compatibility may also depend on the connectivity [165], avoiding to bundle together edges whose nodes are all disconnected in the graph.

When not otherwise stated, **MING** missed neighbours edges are bundled using an adaptation of **FDEB** using the connectivity compatibility and the Lorentzian potential for electromagnetic forces proposed in [165]. The result of that method for the graph of Fig. 4.7a, is presented Fig. 4.7c. This bundling leads to a slight improvement of readability, while avoiding aggregation of incoherent edges. Yet, it suffers from some limitations in terms of scalability (i.e. adaptability to dataset with very high number of points). Indeed, it is not sufficiently strong to make highly cluttered graphs readable, and its computational complexity is $O\left(|E^{\text{rel}}(\kappa)|^2\right)$, with the number of edges of the relevance graph $|E^{\text{rel}}(\kappa)| = O(N\kappa)$.

Meta-Nodes and Bundling by Community

To display large graphs, a possible solution is to aggregate groups of points into meta-nodes and edges between a same pair of those meta-nodes as meta-edges [101]. Directly applying that to **MING** would require to merge several map points and pairwise relations, thus losing some information. Yet, building on the idea behind this approach, we may perform a *bundling by community* for **MING** relevance graph. This bundling first identifies communities of points (i.e. meta-

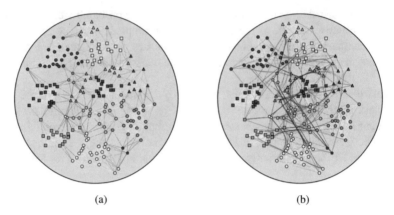

(a)   (b)

Fig. 4.8 MING bundling by community. The sixteen communities obtained by Louvain community detection algorithm [19] on the graph of reliable edges are shown by markers shapes and colours. (**a**) Retrieval graph ($\kappa = 10$). (**b**) Relevance graph ($\kappa = 15$)

nodes) and then strongly bundles together all edges linking the same communities (i.e. belonging to the same meta-edge). As such, the property of belonging to the same meta-edge constitutes a binary measure of compatibility between edges.

For **MING**, the definition of communities must take into account the nodes connectivity in both the data and embedding spaces. Indeed, edges should be bundled together only if their vertices are similar points (connectivity compatibility of relevance edges) and positioned in the same regions of the map (geometric compatibility of relevance edges). Hence, the communities should be reliable map clusters. These may be obtained automatically by performing community detection on the graph of reliable edges, for instance using the Louvain algorithm [19]. This algorithm determines communities by maximizing the modularity, which means maximizing the edges within a same community and minimizing the edges between different communities. The communities obtained for the reliable edges of the retrieval graph with a small value of κ is shown Fig. 4.8. Once the communities are defined, the unreliable edges of the relevance graph belonging to the same macro-edge may be bundled together using **KDEEB** algorithm. This leads to the bundling presented Fig. 4.8 for the relevance graph with a slightly higher value of κ.

4.3.6 Oil Flow

In this section, **MING** is applied to the analysis of the oil flow dataset introduced Sect. 1.1.7. Figure 4.9a presents the repartition of those classes in a 2D map of that dataset. The "stratified" class appears as one cluster occupying the center of the map, while the "annular" class is split in several map clusters distributed around this class. **MING** graphs for that map (see Fig. 4.9b and c) tend to confirm the reliability

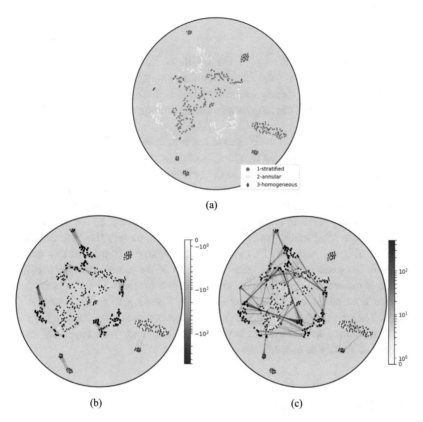

Fig. 4.9 MING with $\kappa = 15$ and MRRE distortion measure, for the oil flow dataset embedded with tSNE [183]. (**a**) Classes. (**b**) Retrieval graph. (**c**) Relevance graph

of these map clusters, and also show some connections between the "stratified" and "annular" classes, at the extremities of the central map cluster. Furthermore, the missed neighbours edges indicate that the different groups of the "annular" class have been torn, which could indicate that in the original space the "stratified" and "annular" classes are both in one component, with a few points of connection between the two classes.

As for the "homogeneous" class, it is represented as many small map clusters mostly separated from the other two classes. Those clusters are well-represented individually, but their positioning with respect to other clusters is not ideal. Indeed, for some points of the "homogeneous" class, the nearest neighbours, on the border of some neighbouring cluster, tend to be false neighbours, while their true neighbours, which are positioned closer to the core of that neighbouring cluster are missed. This characterizes a situation where the high dimensional neighbourhood structure may not be rendered properly in the 2D map, so that the points are moved away.

4.3.7   *COIL-20 Dataset*

MING is then applied to a map of the COIL-20 dataset described Sect. 1.1.7. Figure 4.10 presents the results of MING for a subset of the COIL-20 dataset containing ten out of the twenty classes, and half of the angular steps, in order to improve the readability of the map, where points are represented by their associated image.

The white edges allow to identify reliable continua in the map. This data structure may be explained by the fact that two images of the same object with a slight rotation are often very similar. Since the angles range from 0° to 360°, these continua form cycles, which may be well-represented as for the cat class (zoom I), or torn as for the "piggy bank" class (zoom II). In the case where the object is invariant by rotation, all images are very similar, leading to a cluster-like structure, as for the "cylindric cup" class (zoom V). Finally, some objects are similar, such as the different models of cars. In that case, there may be more similarity between images of two different objects with the same angle than between images of the same object with different angles. This induces the parallel continua observed for the two car classes (zoom III). Those continua cross each other (zoom IV) leading to a slight stretch. The red edges linking together the extremities of those continua also confirm the cyclic structure, whereas some red edges between the "car" and "medicine box" classes show some similarity of shapes between those objects.

4.3.8   *MING Perspectives*

MING constitutes a tool for interpretative support providing local and directional assessment of distortions of the most critical neighbourhood relations. Yet, it suffers from certain limitations. It is especially sensitive to scaling and visual clutter issues, and some technical details such as the grey background which is not ideal for perceiving colours should be improved. Future work could consider the following ameliorations and extensions.

In terms of graph display, the grey background may disturb the eye for the perception of edge colours. This could be corrected by designing a diverging colourmap from red to blue with a central colour that could be visible on a white background such as grey or black, though this may lead to a loss of colour gradient. For edge bundling, the Attribute-Driven Edge Bundling (ADEB) [146] could be well-suited for MING. Indeed, it adapts the density-based approach by introducing a measure of compatibility between edges based on their direction and other attributes. However, there is no public implementation of that algorithm available yet.

For the applications, MING approach could be extended to help interpret higher dimensional embeddings presented with 3D scatter plots or SPLOM. It may also extend to other graph-based indicators using graphs symmetrically defined in both

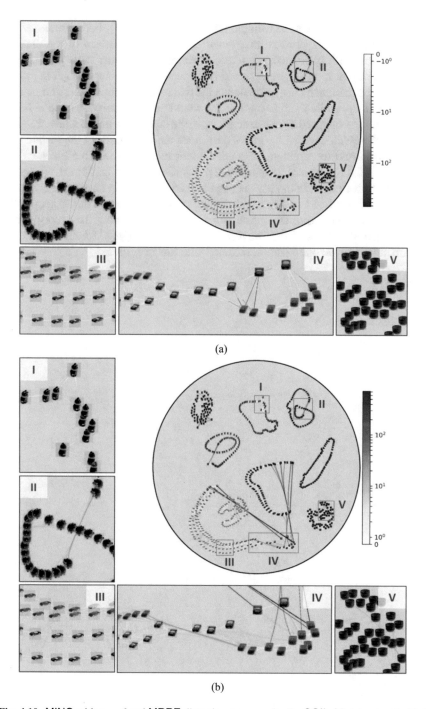

Fig. 4.10 MING with $\kappa = 6$ and MRRE distortion measure, for the COIL-20 dataset embedded with tSNE [183]. (**a**) Retrieval graph. (**b**) Relevance graph

spaces, such as the Extended Minimum Spanning Tree [133]. Furthermore, MING could be used to compare two maps of the same dataset instead of the data and map, as done with other many-to-many interpretative support tools [44, 122]. In that approach, the retrieval graph would be the nearest neighbours graph in the first map, with the reference rank computed in the first map, and image rank computed in the second map, while the relevance graph would be symmetrically defined for the second map. As opposed to the standard case, the equivalent of the relevance graph could be represented with the vertices positioned onto the points of the second map. This would avoid the problem of tangled edges mostly responsible for the visual clutter in MING relevance graph. Finally, MING could be combined with a DR technique to interactively modify the map. Indeed, it may help an analyst to identify specific neighbourhood relations that should be better preserved in the map or that could be distorted with little impact (for instance by selecting the corresponding edge). Then, the stress of the DR technique may be weighted to account for that information, and the optimization resumed with this new stress.

Chapter 5
Stress Functions for Unsupervised Dimensionality Reduction

> *Behold yon miserable creature. That Point is a Being like ourselves, but confined to the non-dimensional Gulf. He is himself his own World, his own Universe; of any other than himself he can form no conception; he knows not Length, nor Breadth, nor Height, for he has had no experience of them; he has no cognizance even of the number Two; nor has he a thought of Plurality, for he is himself his One and All, being really Nothing.*
>
> *Flatland: A Romance of Many Dimensions*
> Edwin Abbot

Dimensionality Reduction (DR) represents a set of points $\{\xi_i\}$ in a high dimensional metric data space $\mathcal{D}$ by associated points $\{x_i\}$ in a low-dimensional embedding space $\mathcal{E}$. This representation defines a mapping $\Phi : \mathcal{D} \longrightarrow \mathcal{E}$ such that $\Phi(\xi_i) = x_i$ for all i. This mapping must preserve as much as possible the structure of data. Chapters 5, 6 and 7 focus on Dimensionality Reduction techniques, which constitute technical solutions for defining the mapping Φ.

DR techniques mostly rely on the minimization of a stress function. The stress defines what must be preserved in priority in the data structure and how to penalize the potential distortions of that underlying structure. Hence, it allows to compare different mappings, in order to determine the best possible mapping according to that stress (minimum of the function). In most cases, the minimum of the stress function is searched using non-convex optimization techniques, while in a few others the global minimum may be obtained directly. Though the performances of a DR method depend on the combination of its stress and its specific optimization process, we choose here to present them as independent components. Thus, Chaps. 5 and 6 describe different stress functions for unsupervised and supervised DR, while the details of non-convex optimization are given in Chap. 7. The latter also addresses possible techniques for reducing the algorithms complexity, as well as out of sample positioning, extending the mapping to points out of the original dataset.

The current Chapter, detailing stress functions of unsupervised DR methods, is organized as follows. First, it presents spectral methods relying on a linear projection, either from the data space or from a kernel space of higher dimensionality

S. Lespinats et al., *Nonlinear Dimensionality Reduction Techniques*, https://doi.org/10.1007/978-3-030-81026-9_5

(Sect. 5.1). Then, it goes into metric and non-metric non-linear Multi-Dimensional Scaling (MDS) (Sect. 5.2), which attempt to preserve either the distances or the neighbourhood ranks. Next, a focus is made on Neighbourhood Embedding methods that aim at preserving the neighbourhood structure characterized by normalized neighbourhood membership degrees (Sect. 5.3). This is distinguished from graph embedding techniques, which consider neighbourhood relations as edges of a graph (Sect. 5.4). Finally, methods defining a mapping based on an artificial neural network are considered (Sect. 5.5).

In terms of notations, we denote ζ the global stress of a DR method. It may also be defined through the point-wise stress ζ_i, aggregated either by sum or average, leading to:

$$\zeta \propto \sum_i \zeta_i.$$

This stress may be minimized either with respect to parameters of an explicitly defined mapping Φ (parametric methods), or with respect to the positions of embedded points $\{x_i\}$ (non-parametric methods). Several methods presented here are implemented in the Python toolbox scikit-learn [142].

5.1   Spectral Projections

Spectral projections methods mostly rely on the idea of performing a multidimensional orthogonal linear projection, either from the data space $\mathcal{D}$ or from an extended feature space $\mathcal{K}$ for kernel methods. This requires that $\mathcal{D}$ and $\mathcal{K}$ be inner product spaces.

In contrast with most other methods for Dimensionality Reduction, spectral projection methods define convex stress functions. Hence, the global optimum of that stress can be obtained with certainty. In the broader sense, a projection is defined as follows:

Definition 5.1 A mapping $\Phi : \mathcal{D} \longrightarrow \mathcal{E}$ is a projection iff $\Phi = b \circ p$, where $p : \mathcal{D} \longrightarrow \mathcal{D}$ is an idempotent transformation (i.e. such that $p \circ p = p$) and $b : p(\mathcal{D}) \to \mathcal{E}$ is a bijection.

It is linear, if both b and p are linear.

5.1.1   Principal Component Analysis

Principal Component Analysis (PCA) [94, 141] is an historic dimensionality reduction method, stemming from 1901, that is still one of the best-known today.

Fig. 5.1 Example of PCA applied to a two-dimensional Gaussian dataset (in blue). Points are projected along direction a_2 onto the linear subspace defined by vector a_1, which leads to the red projected points

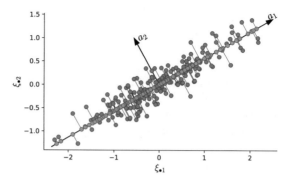

It applies to a set of data points Ξ in a Euclidean space $\mathcal{D} = (\mathbb{R}^\delta, || \cdot ||_2)$ the affine projection of rank d (illustrated Fig. 5.1) that:

- preserves the most of the data variance (Sect. 5.1.1.1),
- is the closest from the original data (Sect. 5.1.1.2),
- provides the most likely parameters of a latent variable model (Sect. 5.1.1.3).

This wide variety of interpretations induces many possible applications and has lead to the development of many variants, as detailed below. The best *affine* projection for those criteria is the best *linear* projection of the centred data $\tilde{\Xi} = \Xi - \frac{1_{N \times N}}{N}\Xi$ (see Annex D.4).

As a linear projection, PCA finds two matrices X and A. The matrix $X \in \mathbb{R}^{N \times d}$ is the expected output of a DR method, whose rows $\{x_i\}$ are the coordinates of the embedded points in the embedding space $\mathcal{E} = (\mathbb{R}^d, || \cdot ||_2)$. Conversely, the matrix $A \in \mathbb{R}^{d \times \delta}$ defines, with its row vectors $\{a_i\}$, an orthonormal basis of subspace $p(\mathcal{D}) \subset \mathcal{D}$ onto which the data are projected (where p is the idempotent transformation from Definition 5.1). The space $p(\mathcal{D})$ is in bijection with the embedding space $\mathcal{E}$, and the product XA gives the coordinates of the projected points X in the data space coordinates.

5.1.1.1   Variance Interpretation

PCA performs the projection best preserving the variance of the dataset. That is:

$$\zeta^{\text{PCA}} = ||C_{\tilde{\Xi}} - C_{XA}||_2 \tag{5.1}$$

with $C_{\tilde{\Xi}} = \frac{1}{N}\tilde{\Xi}^\top\tilde{\Xi}$ being the covariance matrix of the centred matrix $\tilde{\Xi}$ of size $N \times \delta$ and $|| \cdot ||_2$ the Frobenius matrix norm [201]. This minimization problem may be solved by eigenvalue decomposition of the data covariance matrix see Annex D.5). The projection obtained that way maximizes the amount of variance in the embedding space. Moreover, the comparison of the total amount of variance between the data and embedding spaces provides an estimation of information loss induced by the projection. As such, it provides a measure of the faithfulness of that specific representation.

5.1.1.2 Reconstruction Error

PCA finds the projection of the data that is the closest to the data points in the sense of the root mean square error:

$$\zeta^{\text{PCA}} = ||\tilde{\Xi} - XA||_2, \tag{5.2}$$

with $|| \cdot ||_2$ the Frobenius norm (equivalent for matrices of the Euclidean norm for vectors). This formulation may lead to solve the problem by Singular Value Decomposition (SVD) (see Annex D.1) [179].

This projection may be interpreted as a low-rank linear model of the data of the form $\tilde{\Xi} \approx XA$ [179]. In that framework, the stress from Eq. (5.2) is the *reconstruction error* of the model. This relates to linear regression, but where both the model parameters A and explanatory variables X are unknown. As such, the problem can be solved by alternate minimization with respect to X and A. Since $\tilde{\Xi} \approx XA$ is an approximate matrix factorization of the data matrix of size $N \times \delta$, with two components of size $N \times d$ and $d \times \delta$, this finds direct applications in data compression [104].

In this low-rank model approach, the data points are approximated by linear combinations of d archetypes $\{a_i\}$. These *archetypes* represent an idealized profile among the data points, characteristic of some tendency [179].

To increase the interpretability of such representation, constraints of sparsity or positivity may be added for matrices X and A. Sparse Principal Component Analysis (SPCA) [211] searches to obtain a sparse matrix A, so that each archetypical direction be a combination of only a few data features. Non-negative Matrix Factorization (NMF) [104] imposes that all elements of X and A be positive. For X, this implies that all linear combination are additive (no substration of archetypes). For A, this means that all elements of A have positive components, which is more meaningful when the coordinates of data points are positive by construction (e.g., images or spectra).

5.1.1.3 Latent Variable Model

Probabilistic PCA [176] proposes another interpretation of PCA considering that the random variable ξ from which the data points are drawn may be explained by latent variable model. This model assumes that the data manifold is a low-dimensional affine subspace. It may be expressed as:

$$\xi = \bar{\xi} + xA + \epsilon, \tag{5.3}$$

with $\bar{\xi} + xA$ being the expression of an affine transform of the low-dimensional latent variable x, and ϵ an isotropic high dimensional Gaussian noise.

With these hypotheses, PCA returns the most likely values for the parameters $\bar{\xi}$ and A of the affine transform. This interpretation leads to the use of the Expectation Maximization algorithm for finding those parameters. It also justifies the use of PCA for noise filtering.

Exponential family PCA [37] extends this approach to other types of distributions of noise ϵ belonging to the exponential family. The exponential family regroup the Gaussian, Poisson and Bernoulli distribution which may respectively model noise for real, integer and binary variables. In that general case, the most likely manifold is not necessary affine and the latent variable model from Eq. (5.3) may be rewritten as:

$$\xi = \phi(xA) + \epsilon,$$

with ϕ a non-linear link function. The measure of error of Eq. (5.2) is also adapted, replacing the squared Euclidean norm by another divergence $\mathscr{D}$ in the family of Bregman divergences:

$$\zeta^{\text{gPCA}} \triangleq \sum_i \mathscr{D}\left(\xi_i, \phi(x_i A)\right).$$

Both the link function and the divergence are derived from the probability density function of the considered distribution. We may note that this extension of PCA relates to the extension of linear regression by Generalized Linear Mixed Models.

5.1.2 Classical MultiDimensional Scaling

The classical Multi-Dimensional Scaling (cMDS) [177] finds, for data points in a Euclidean space, the linear projection that best preserves the inner product between the vectors (with the data centroid as origin). This translates as:

$$\zeta^{\text{cMDS}} \triangleq ||\Gamma_{\tilde{\Xi}} - \Gamma_{XA}||_2, \tag{5.4}$$

with $\Gamma_{\tilde{\Xi}} = \Xi \Xi^{\top}$ the Gram matrix of the matrix Ξ, whose element (i, j) is the dot product $\langle \xi_i, \xi_j \rangle$ between the ith and j^{th} row vector of $\tilde{\Xi}$, and $|| \cdot ||_2$ the Frobenius norm [201]. This expression is dual to the formulation of PCA in terms of covariance, presented by Eq. (5.1). As such, it may be shown to give similar results to PCA [80].

Furthermore, cMDS also extends to any dataset Ξ in a metric space. This relies on the interesting property that for any distance matrix Δ_{Ξ}, there exists a centred set of points $\tilde{\Xi}'$ in a high-dimensional Euclidean space, such that $\Delta_{\tilde{\Xi}'} = \Delta_{\Xi}$ [206].

The centred Gram matrix $\mathbf{\Gamma}_{\tilde{\Xi}'}$ of that hypothetical dataset is given by the following transformation:

$$\mathbf{\Gamma}_{\tilde{\Xi}'} \triangleq -\frac{1}{2}\left(\mathbf{I}_{N\times N} - \frac{\mathbf{1}_{N\times N}}{N}\right)\mathbf{\Delta}_{\Xi} \circ \mathbf{\Delta}_{\Xi}\left(\mathbf{I}_{N\times N} - \frac{\mathbf{1}_{N\times N}}{N}\right), \qquad (5.5)$$

with $\circ$ the Hadamard product. Hence, in that general case cMDS is equivalent to PCA applied to the hypothetical Euclidean dataset $\tilde{\Xi}'$.

5.1.2.1   Limitations of Linear Methods

Linear methods rely on the assumption that the data manifold is a linear (or affine) subspace. This means that there exist linear relations between the data features. Yet, in many real cases, some relations are non-linear, leading to non-linear manifolds. Figure 5.3 illustrates the inadequacy of linear methods for embedding the non-linear Swiss roll manifold. In that case, the data structure is highly distorted with a lot of manifold gluings. Indeed, the linear projection squashes the spiral structure while a more intuitive solution would unfold it.

5.1.3   Kernel Methods: Isompap, KPCA, LE

In order to unfold a non-linear manifold, other non-linear spectral methods have been developed, among which Isometric feature mapping (Isomap) [175], Locally Linear Embedding (LLE) [154] or Laplacian Eigenmaps [14]. All these methods may be formulated as sub-cases of KPCA as shown in [84]. Diffusion maps [34] also fall into this framework.

5.1.3.1   Kernel PCA

Kernel Principal Component Analysis (KPCA) [159] actually performs a cMDS in an extended feature space in order to benefit from the blessing of dimensionality. This extended feature space $\mathcal{K}$ is equipped with an inner product $\langle \cdot, \cdot \rangle_{\mathcal{K}}$. To avoid intractable computations of dot products in this very high dimensional space, KPCA uses the kernel trick (as explained in [28]).

This computational trick may be used to extend any given algorithm relying solely on the computation of inner products to every extended feature spaces respecting the following conditions:

- There exists a feature transformation $\Psi : \mathcal{D} \longrightarrow \mathcal{K}$, mapping data points to points in an extended feature space $\mathcal{K}$ of higher dimensionality.
- There exists a kernel function $\gamma_{\Psi} : \mathcal{D} \times \mathcal{D} \longrightarrow \mathbb{R}$ on the data space, such that:

$$\gamma_{\Psi}(\xi_i, \xi_j) \triangleq \langle \Psi(\xi_i), \Psi(\xi_j) \rangle_{\mathcal{K}}.$$

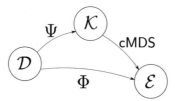

Fig. 5.2 Spaces involved in the kernel based methods. The data points from the data space $\mathcal{D}$ are theoretically mapped to a inner product space $\mathcal{K}$ using a mapping Ψ. Then, a projection is performed in that space using cMDS. In practice, effective computations in the space $\mathcal{K}$ are avoided using the kernel trick

In these cases, the inner product may be computed directly in $\mathcal{D}$ using the kernel function, rather than in the much higher dimensional space $\mathcal{K}$ (Fig. 5.2).

For kernel **PCA**, the centred Gram matrix in $\mathcal{K}$ is thus given by:

$$\tilde{\boldsymbol{\Gamma}}_{\Psi(\Xi)} \triangleq \left(\boldsymbol{I}_{N \times N} - \frac{\boldsymbol{1}_{N \times N}}{N} \right) \boldsymbol{\Gamma}_{\Psi(\Xi)} \left(\boldsymbol{I}_{N \times N} - \frac{\boldsymbol{1}_{N \times N}}{N} \right),$$

where $\boldsymbol{\Gamma}_{\Psi(\Xi)}$ is the kernel matrix (and Gram matrix in the extended space) whose element (i, j) is $\gamma_\Psi(\xi_i, \xi_j)$. The coordinates of points in the embedding space may then be obtained as in **cMDS** (see Sect. 5.1.2).

5.1.3.2 Isomap

The **Isomap** algorithm [175] attempts to perform **cMDS** directly in the data manifold $\mathcal{M}$ equipped with its geodesic distance, approximated by the shortest path distances in a nearest neighbours graph. Hence, the centred Gram matrix of the points $\{\Psi(\xi_i)\}$ in the extended feature space $\mathcal{K}$ is obtained from the matrix of geodesic distances using Eq. (5.5). Conversely to **PCA**, **Isomap** is able to unfold the manifold structure for the simple **Swiss roll** dataset as shown Fig. 5.3c.

5.1.3.3 Laplacian Eigenmap

Laplacian Eigenmap (**LE**) [14] is fundamentally a graph embedding technique. For general metric data, it first builds a nearest neighbours graph, weighted by pairs of points similarities (computed with a Gaussian kernel). It attempts to minimize the distances between embedded points that are strongly connected in the graph. This yields the following stress:

$$\zeta^{\text{LE}} \triangleq \sum_i \sum_{j \neq i} w_{ij} ||x_i - x_j||_2^2, \tag{5.6}$$

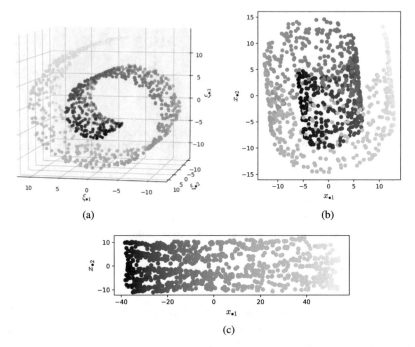

Fig. 5.3 Projection of the Swiss roll non-linear manifold introduced Sect. 1.1.7 by PCA and Isomap. The main curvilinear component (curvilinear abscissa along the spiral) is encoded in colour. With the linear projection of PCA, the structure is squashed, placing close from each other points that are very distant along the manifold. With the non-linear Isomap, the structure is unrolled, clearly presenting the curvilinear components parametrizing the manifold. (**a**) 3D dataset. (**b**) PCA projection. (**c**) Isomap embedding

with $\{w_{ij}\}$ the weights of the graph. To avoid reaching the trivial minimum embedding all points at the same position, the following covariance constraint is added:

$$X^\top \text{diag}\,(k_{11}, \ldots, k_{ii}, \ldots, k_{NN})\,X = I_{d\times d}, \qquad (5.7)$$

with diag defining a diagonal matrix based on its diagonal elements and k_{ii} the in-degree of the node i given by:

$$k_{ii} \triangleq \sum_{k\neq i} w_{ki}.$$

This ensures that nodes with highest degree are placed further away from the centroid of embedded points.

The constrained optimization problem defined by Eqs. (5.6) and (5.7) may be solved as a generalized eigenvalue problem involving the graph Laplacian. It is equivalent to performing cMDS on the matrix of commute time distances in the graph [149].

5.1.3.4 Locally Linear Embedding

Locally Linear Embedding (LLE) [154] assumes that small neighbourhood in the manifold are locally linear. Based on that, each point may be reconstructed as a linear combination of its neighbours. First, LLE computes the weights $\{w_{ij}\}$ that minimize the reconstruction error:

$$\zeta^{\text{LLE}^\star} \triangleq \sum_i \left\| \xi_i - \sum_{j \neq i} w_{ij} \xi_j \right\|_2^2.$$

This is solved with the constraints that $w_{ij} = 0$ if $j \notin v_i(\kappa)$ (only a combination of the κ nearest neighbours) and $\sum_{j \neq i} w_{ij} = 1$ (ensuring the invariance of weights to translation of the data points). In a locally linear embedding, the points should remain the same linear combination of their neighbours. Hence, the second phase of LLE consists in minimizing the reconstruction error with respect to the embedd points positions and with fixed weights $\{w_{ij}\}$, that is:

$$\zeta^{\text{LLE}} \triangleq \sum_i \left\| x_i - \sum_{j \neq i} w_{ij} x_j \right\|_2^2.$$

5.2 Non-Linear MultiDimensional Scaling

As opposed to spectral projection methods, non-linear MultiDimensional Scaling (MDS) methods are non-parametric. This means that the map is obtained by directly optimizing a stress function with respect to the points coordinates. The main goal expressed by MDS stress functions is to reflect as faithfully as possible the pairwise distances between points in the data space by their corresponding pairwise distances in the embedding space. Hence, contrary, for example, to PCA, the extracted features, namely the axes of the map, have no direct meaning, and the map may be rotated, symmetrized or translated without changing its meaning.

We may distinguish between metric and non-metric MDS, which focus on preserving either the values of distances or the order of distances. Indeed, due to the phenomenon of norm concentration occurring for high dimensional data (see Sect. 2.1), the distances values may not be identical in both spaces. Thus, non-metric MDS consider (directly or indirectly) neighbourhood ranks rather than distances. For those methods, the map may also be scaled without impact on its meaning. Furthermore, most MDS stresses tend to prioritize the preservation of short distances or small ranks. This is based on the assumption that inter-point distances measured directly in the high dimensional ambient space are closer from distances along the lower dimensional manifold for smaller scales.

5.2.1 Metric MultiDimensional Scaling

The most basic cost function used to optimize a map in MDS is the raw stress [21]:

$$\zeta_i \triangleq \sum_j (\Delta_{ij} - D_{ij})^2. \tag{5.8}$$

This is opposed to the strain [21], which is another name of the stress function of classical MDS, namely a reformulation of the matrix Eq. (5.4):

$$\zeta_i \triangleq \sum_j (\langle \xi_i, \xi_j \rangle - \langle x_i, x_j \rangle)^2.$$

Potential of Heat-diffusion for Affinity-based Transition Embedding (**PHATE**) [130] rely on an equivalent of the raw stress of Eq. (5.8) with diffusion distances detailed for network data (see Sect. 1.1.6).

5.2.1.1 Sammon Non-Linear Mapping and Curvilinear Component Analysis

The raw stress penalizes equally the distortion of all distances. Yet, assuming that data live in a non-linear manifold, the small distances are considered more important. Indeed, they approximately correspond to distances along the manifold, while higher distances may need to be distorted to allow manifold unfolding. Hence, weights may be introduced to ponder the distances errors accordingly. Sammon Non Linear Mapping (**SNLM**) [158] and Curvilinear Component Analysis (**CCA**) [54] are two closely defined MDS methods based on this idea. Their stress functions are given respectively by:

$$\zeta_i^{\text{SNLM}} \triangleq \sum_{j \neq i} (\Delta_{ij} - D_{ij})^2 w \left(\Delta_{ij} \right), \tag{5.9}$$

$$\zeta_i^{\text{CCA}} \triangleq \sum_{j \neq i} (\Delta_{ij} - D_{ij})^2 w \left(D_{ij} \right), \tag{5.10}$$

with w a decreasing function of distances (e.g., the reciprocal function $x \mapsto 1/x$).

These stresses mainly penalize important distances errors associated with high weights. Hence, **SNLM** mostly prevents manifold tears ($D_{ij} \gg \Delta_{ij}$), namely errors of representation of small data space distances. Yet, high distances may be stretched in order to unfold the manifold, with a moderate impact on the stress value. Conversely, **CCA** mostly avoids manifold gluing ($D_{ij} \ll \Delta_{ij}$), which are distance errors for which the distance in the embedding space is small [114, 185]. Another method called Curvilinear Distance Analysis (**CDA**) [108], applies **CCA** stress to estimated geodesic distances (using the method proposed in [175]). This use of geodesic distances tends to cause a stretch of the long distances in the data space, which favours manifold unfolding.

5.2.1.2 Local MultiDimensional Scaling

Local Multi-Dimensional Scaling (LMDS) [187] combines linearly the stresses of CCA and SNLM, balancing their influence through a trade-off parameter τ. Thus, its stress may be written as:

$$\zeta_i^{\text{LMDS}} \triangleq \sum_{j \neq i} (\Delta_{ij} - D_{ij})^2 w_{ij}(\tau), \tag{5.11}$$

where the weights $\{w_{ij}(\tau)\}$ are defined as:

$$w_{ij}(\tau) \triangleq \tau w \left(\frac{\Delta_{ij}}{\sigma_i} \right) + (1 - \tau) w \left(\frac{D_{ij}}{s_i} \right),$$

with $\tau \in [0, 1]$ the trade-off parameter, σ_i and s_i scale parameters associated to point i in the data and embedding spaces. In LMDS, w is a flat kernel, i.e. the indicator function of $[-1, 1]$, and scale parameters are both set as the distance between i and its κth nearest neighbour in the data space.

Changing the value of τ allows the user to tune the trade-off between the two types of distortions, with high values of τ penalizing more manifold tears, and low values of τ penalizing more manifold gluing. For extreme cases $\tau = 1$ and $\tau = 0$, LMDS stress is respectively equivalent to those of SNLM and CCA.

5.2.1.3 Data-Driven High Dimensional Scaling

Data-Driven High Dimensional Scaling (DD-HDS) [114] uses a weighting strategy penalizing both false and missed neighbours and also accounting for the curse of dimensionality. It is based on the following stress function:

$$\zeta_i \triangleq \sum_i |\Delta_{ij} - D_{ij}| w \left(\min(\Delta_{ij}, D_{ij}) \right),$$

where $w(u) = 1 - \text{CDF}(u)$ with CDF denoting the cumulative distribution function of the normal distribution whose parameters are set depending on the distribution of all distances Δ_{ij} in the data space. Assuming that the values of distances should be close, the use of the Gaussian weighting, which is some kind of sigmoid, allows to discriminate efficiently small and high distances. The decrease of weights with the minimum of data and map distances leads to penalize less the distances errors for pairs of points that are far from each other in both spaces.

5.2.2   Non-Metric MultiDimensional Scaling

In non-metric MDS, distances between embedded points are not supposed to correspond directly to the values of data space distances, but rather to reflect their order. In Kruskal algorithm [99], this idea is translated by maximizing the monotonicity of the Shepard diagram plotting embedding space distances against data space distances (as detailed Sect. 3.2.6.1). To assess this monotonicity at a given step of the algorithm, an isotonic regression [100] is applied to fit a monotonic model to the graph of distances $\{D_{ij}\}$ as a function of $\{\Delta_{ij}\}$. This model allows to estimate ideal values $\{\widehat{D}_{ij}\}$ for embedding distances, that are ordered as the distances $\{\Delta_{ij}\}$. Then, the algorithm may optimize the positions of the embedded points in order to minimize the error:

$$\zeta \triangleq \frac{\sum_i \sum_{j \neq i} (\widehat{D}_{ij} - D_{ij})^2}{\sum_i \sum_{j \neq i} D_{ij}^2}$$

The normalization by the denominator is meant to render this stress function invariant to uniform scaling of the embedding, since overall scale is not relevant in this non-metric approach. As opposed to neighbourhood ranks, the order of distances is not computed independently for each point i, but rather for all pairwise distances $\{D_{ij}\}$.

5.2.2.1   RankVisu

RankVisu [115] directly ensures the preservation of neighbourhood ranks, relying on the following pointwise stress:

$$\zeta_i \triangleq \sum_{j \in v_i(\kappa) \cup n_i(\kappa)} (\kappa + 1 - \min(\rho_{ij}, r_{ij}))^+ |\rho_{ij} - r_{ij}|, \tag{5.12}$$

where $\{\rho_{ij}\}$ and $\{r_{ij}\}$ are the neighbourhood ranks, $v_i(\kappa)$ and $n_i(\kappa)$ the κ-neighbourhoods of point i, both in the data and embedding spaces, and $\bullet^+$ the positive part. This stress solely penalizes κ false and missed neighbours, with a higher weighting for the most severe of those distortions. Yet, computationally, this expression in terms of neighbourhood ranks suffers from several limitations. Indeed, embedding space ranks in the map only depend on the relative order of distances. As such, they do not define a global scale. Furthermore, they are piecewise constant functions of the distances, thus leading to a difficult optimization process. Finally, focusing solely on the ranks may lead to a map with imperceptible differences of distances, perceived by the machine as significantly different ranks. To account for these limitations, the stress defined Eq. (5.12) is combined in the early stages of the optimization process with another stress comparing the values of map distances $\{D_{ij}\}$ with those of their associated data space ranks $\{\rho_{ij}\}$ and $\{\rho_{ji}\}$.

5.3   Neighbourhood Embedding Methods

Neighbourhood Embedding techniques correspond here to a family of methods derived from Stochastic Neighbour Embedding (SNE) [87], and also comprising t-distributed Stochastic Neighbour Embedding (tSNE) [183], Neighbourhood Retrieval Visualizer (NeRV) [188, 189] and Jensen Shannon Embedding (JSE) [109]. All these methods are based on the preservation between the data and embedding spaces of soft neighbourhoods. As such, they may be interpreted as tools to perform the task of visual neighbourhood retrieval [189]. This means that the nearest neighbours of each point i, considered as the neighbours retrieved by the analyst, should correspond as well as possible to the nearest neighbours of the same point i in the data space, named relevant neighbours.

Similarly to precision and recall distortion indicators, neighbourhood embedding stress functions account for the differences between the data and embedding space neighbourhoods of each point. Yet, their approach relies on soft neighbourhoods rather than hard neighbourhoods. With hard neighbourhoods, a point may either be inside or outside the neighbourhood, so that the neighbourhood belonging is a binary notion. Conversely, for soft neighbourhoods, this belonging may be quantified with varying levels. As a result, the stress is a smooth function of the position of points in the embedding space, allowing to perform optimization of that stress.

5.3.1   General Principle: SNE

In order to assess the preservation of the neighbourhood structure with respect to the mapping, neighbourhood embedding algorithms measure for each point i, the belonging of any other point j to its soft neighbourhood in both spaces. This gives the neighbourhood membership degrees $\beta_i \triangleq \{\beta_{ij}\}_{j \neq i}$ in the data space and $b_i \triangleq \{b_{ij}\}_{j \neq i}$ in the embedding space. In the original method SNE [87], these membership degrees are defined using Gaussian kernels as:

$$\beta_{ij} \triangleq \frac{\exp\left(-\frac{\Delta_{ij}^2}{2\sigma_i^2}\right)}{\sum_{k \neq i} \exp\left(-\frac{\Delta_{ik}^2}{2\sigma_i^2}\right)} \quad \text{and} \quad b_{ij} \triangleq \frac{\exp\left(-\frac{D_{ij}^2}{2s_i^2}\right)}{\sum_{k \neq i} \exp\left(-\frac{D_{ik}^2}{2s_i^2}\right)}, \tag{5.13}$$

where $\{\sigma_i\}$ and $\{s_i\}$ are scale parameters in the data and embedding spaces. In the neighbourhood retrieval perspective, these normalized neighbourhood membership degree are sometimes interpreted as a probability of selecting a neighbour j in the neighbourhood of a given point i [189].

This softmin formulation benefits from the shift-invariance property [107], allowing to mitigate the phenomenon of norm concentration (see Sect. 2.1). This property may partly justify the better suitability of neighbourhood embedding methods for very high dimensional data (compared with MDS).

The neighbourhood membership degrees being normalized (i.e. summing to one), they behave as discrete probability distributions. Hence, they may be compared between the two spaces using the Kullback–Leibler divergence, leading to a point-wise stress:

$$\zeta_i \triangleq \mathscr{D}_{KL}(\beta_i, b_i), \tag{5.14}$$

with the classical Kullback–Leibler divergence defined by:

$$\mathscr{D}_{KL}(\beta_i, b_i) \triangleq \sum_{j \neq i} \beta_{ij} \log\left(\frac{\beta_{ij}}{b_{ij}}\right). \tag{5.15}$$

Variants of SNE mostly adapt the kernels used to define the membership degrees (Sects. 5.3.5 and 5.3.6) or the divergence measure (Sect. 5.3.3). The kernels determine the link from distances to membership degrees, while the divergence characterizes the comparison of membership degrees between the two spaces. Moreover, the link between distances and membership degrees is influenced by the scale parameter defining the size of the neighbourhood to preserve (see Sect. 5.3.2).

5.3.2   Scale Setting

The data space scale parameter σ_i is set for each point i so that the perplexity κ_i of distribution of membership degrees β_i reaches a given value $\kappa^\star$. The perplexity is defined by:

$$\kappa_i \triangleq \exp(H_i),$$

with the entropy H_i of the distribution β_i given by:

$$H_i \triangleq -\sum_{j \neq i} \beta_{ij} \log \beta_{ij}.$$

Hence, setting the scale parameter σ_i by the perplexity is equivalent to imposing that the entropy H_i be equal to:

$$H_i = \log \kappa^\star. \tag{5.16}$$

membership degrees with scale set in that way may be qualified as *isentropic* [194].

The perplexity of a probability distribution roughly corresponds to an equivalent number of possible outcomes. Thus, this is the soft equivalent of setting the scale

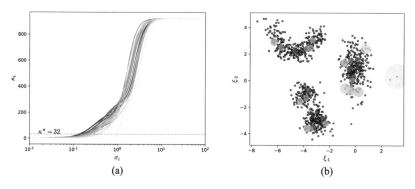

Fig. 5.4 Evolution of the perplexity as a function of the scale parameter for the neighbourhoods of all points of the toy dataset used in Sect. 1.2 (see Fig. 1.7). The scales are shown as the radius of circles around the corresponding points in the left panel. (**a**) Perplexity against scale. (**b**) Scale obtained for fixed perplexity

of the region of interest by a number of neighbours as in **LMDS** (see Sect. 5.2.1.2). Indeed, considering hard neighbourhoods defined by replacing the Gaussian kernels in Eq. (5.13) by flat kernels, the perplexity is the number of neighbours closer than σ_i (see Annex B.1.1).

The root-finding problem defined by Eq. (5.16) is usually solved by the bisection or Newton methods [194]. It is illustrated in Fig. 5.4, presenting the perplexity as a function of the scale parameter. The perplexity increases with the scale. It ranges from 1 (for $\sigma_i \rightarrow 0$) to $N - 1$ (for $\sigma_i \rightarrow +\infty$). Indeed, if the neighbourhood size is infinitely small, only the closest neighbours are considered. Inversely, for an infinitely high scale, all points are equivalent neighbours.

The embedding space scale parameter is either set equal to the data space parameter $s_i = \sigma_i$ [189] leading to a nearly metric behaviour, or uses a uniform value, such as $s_i = 1$ [87, 183], leading to a non-metric behaviour.

5.3.3 *Divergence Choice: NeRV and JSE*

The classical Kullback–Leibler divergence $\mathscr{D}_{KL}(\beta_i, b_i)$, defined in Eq. (5.15) measures the difference between probability distributions β_i and b_i. This divergence being asymmetrical, its two dual formulations may be used to compare the neighbourhood membership degrees [189] leading to:

- soft recall $\mathscr{D}_{KL}(\beta_i, b_i)$ penalizing more the missed neighbours,
- soft precision $\mathscr{D}_{KL}(b_i, \beta_i)$, penalizing more the false neighbours.

Due to this formulation, soft precision penalizes more false neighbours (errors with high b_{ij}), whereas soft recall penalizes more missed neighbours (errors with high

β_{ij}). These two terms relate to precision and recall quality indicators when the soft neighbourhoods tend towards hard neighbourhoods [189], as shown in Annex B.2.

In order to provide a tunable trade-off between these two terms, two types of mixtures are proposed in the literature. Neighbourhood Retrieval Visualizer (NeRV) [188, 189] relies on the following stress:

$$\zeta_i^{\text{NeRV}} \triangleq \mathcal{M}_{\text{KL},1}(\beta_i, b_i, \tau),$$

where $\mathcal{M}_{\text{KL},1}(\beta_i, b_i, \tau)$ is the type 1 mixture with trade-off parameter $\tau \in [0, 1]$ defined by:

$$\mathcal{M}_{\text{KL},1}(\beta_i, b_i, \tau) \triangleq \tau \mathcal{D}_{\text{KL}}(\beta_i, b_i) + (1 - \tau) \mathcal{D}_{\text{KL}}(b_i, \beta_i). \tag{5.17}$$

Similarly, Jensen Shanon Embedding (JSE) [109] relies on the stress:

$$\zeta_i^{\text{JSE}} \triangleq \mathcal{M}_{\text{KL},2}(\beta_i, b_i, \tau),$$

with $\mathcal{M}_{\text{KL},2}(\beta_i, b_i, \tau)$ the type 2 mixture defined by:

$$\mathcal{M}_{\text{KL},2}(\beta_i, b_i, \tau) \triangleq \frac{1}{\tau} \mathcal{D}_{\text{KL}}(\beta_i, \mu_i) + \frac{1}{1 - \tau} \mathcal{D}_{\text{KL}}(b_i, \mu_i), \tag{5.18}$$

where $\mu_i \triangleq (\mu_{ij})_{j \neq i}$ with $\mu_{ij} \triangleq \tau b_{ij} + (1 - \tau)\beta_{ij}$. In the following, when both mixtures are interchangeable they will be denoted as $\mathcal{M}_{\text{KL},12}(\beta_i, b_i, \tau)$. Those two mixtures converge to soft recall (which is the stress of SNE) for $\tau \to 1$ and to soft precision for $\tau \to 0$.

For $\tau \in]0, 1[$, the two mixtures lead to different behaviours. In terms of optimization, the type 1 mixture (NeRV) has an elastic behaviour, while the type 2 mixture (JSE) may lead to plastic elongations [107] (see Sect. 7.1.5.1). In addition, NeRV has a far higher difference of penalization between mild and severe distortions than JSE. Indeed, a severe distortion corresponds to a case where β_{ij} or b_{ij} is close to 0 while its counter-part is not. Thus, the log term tends to ∞ for the type 1 mixture, while it remains bounded for the type 2. As such, NeRV would theoretically lead to better trustworthiness and continuity values (those indicators accounting for distortions severity), while JSE would lead to better precision and recall values. We only present here solutions relying on the Kullback–Leibler divergence, but many different divergences may be considered [204].

5.3.4 Symmetrization

Due to the definition of membership degrees given by Eq. (5.13), the derivatives of the stress function of Eq. (5.14) with respect to distances D_{ij} and D_{ji} are not equal. This leads to a rather complex formulation of the gradient of this stress function (see

Annex C). To simplify the computation of the gradient, a symmetrized expression of membership degrees may be employed [41, 183].

In the embedding space, this symmetrization is obtained by normalizing over all pairs of points. This gives the symmetrized membership degrees b'_{ij}:

$$b'_{ij} \triangleq \frac{g\left(\frac{D_{ij}}{s}\right)}{\sum_k \sum_{l \neq k} g\left(\frac{D_{ij}}{s}\right)}, \tag{5.19}$$

with g the kernel in the embedding space and s a uniform scaling parameter set to $s = 1$.

For the data space, a simple approach could be to use a similar expression to that of Eq. (5.19), with uniform scale σ [41]. Yet, in that case, membership degrees associated to outliers (i.e. points at a great distance of all other points) would all be close to zero (negligible compared with those of points in denser regions), preventing a good embedding of these outliers. This observation leads to the following definition of symmetrized membership degrees in the data space [183] (employed by most methods derived from tSNE [47, 181, 202]):

$$\beta'_{ij} \triangleq \frac{\beta_{ij} + \beta_{ji}}{2N}. \tag{5.20}$$

It is thus obtained as an arithmetic mean of symmetric terms β_{ij} and β_{ji}, followed by a normalization other all pairs (i, j), so that $\sum_i \sum_{j \neq i} \beta'_{ij} = 1$ (that is the division by N). In order to enable outliers, the arithmetic mean $(\beta_{ij} + \beta_{ji})/2$ may be replaced by a geometric mean $\sqrt{\beta_{ij}\beta_{ji}}$ [161]. Indeed, with the arithmetic mean, a symmetrized membership degree β_{ij} is high if at least one of their non-symmetrized counterparts β_{ij} and β_{ji} is high, leading to representing x_i and x_j close in the embedding. Conversely, with the geometric mean, a high β'_{ij} requires that both β_{ij} and β_{ji} be high, namely that the points are both close neighbours of each other.

With those symmetrized membership degrees, the stress becomes a divergence over all possible pairs (i, j):

$$\zeta' \triangleq \sum_i \sum_{j \neq i} \beta'_{ij} \log\left(\frac{\beta'_{ij}}{b'_{ij}}\right).$$

5.3.5 Solving the Crowding Problem: tSNE

The *crowding problem* [183] corresponds to the fact that the number of neighbours of a point increases significantly faster with the neighbourhood radius in a high-dimensional space than in a low-dimensional one. Hence, assuming that the neighbours at a small distance from the point may be well-represented in the low-

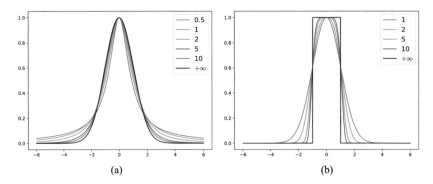

Fig. 5.5 Student and power kernel families for several values of the degree of freedom parameter, with the Gaussian and flat kernels as their limits when the degree of freedom tends to $+\infty$. (**a**) Student kernels. (**b**) Power kernels

dimensional space, there will be far more neighbours at an intermediate distance from the point in the high dimensional data space than what can fit at an intermediate distance in the low-dimensional subspace (without collapsing them together). To tackle that issue, t-distributed Stochastic Neighbour Embedding (tSNE) [183] allows points at an intermediate or high distance from a point to be stretched. This is obtained by using a Student kernel with one degree of freedom in the embedding space, leading to the following expression for membership degrees (with β_{ij} subsequently symmetrized using Eq. (5.20)):

$$\beta_{ij} \triangleq \frac{\exp\left(-\frac{\Delta_{ij}^2}{2\sigma_i^2}\right)}{\sum_{j\neq i} \exp\left(-\frac{\Delta_{ik}^2}{2\sigma_i^2}\right)} \quad \text{and} \quad b'_{ij} \triangleq \frac{(1 + D_{ij}^2)^{-1}}{\sum_k \sum_{l\neq k}(1 + D_{kl}^2)^{-1}}. \tag{5.21}$$

The Student kernel has a heavier tail than the Gaussian kernel (see Fig. 5.5a). Hence, the membership degrees b_{ij} may remain relatively high for very long distances D_{ij}, and thus match their corresponding β_{ij}, allowing for significant stretching. As such, tSNE may be seen as converting *dimensional complexity* into *spatial spreading* of the data.

5.3.6 Kernel Choice

tSNE assumes a very significant difference of dimensionality between the data and the embedding, and induces a level of stretching in coherence with that assumption. Indeed, the Gaussian kernel is the limit of the Student kernel family when the number of degrees of freedom tends to infinity. Yet, for data of low intrinsic dimensionality and/or embedding space of intermediate dimensionality (as may be used for applications other than data visualization) that hypothesis does not hold. Thereby, the discrepancy of kernels should be adapted to the effective magnitude of the crowding problem, in order to avoid over-stretching of the data.

Many methods have attempted to some extent to adapt the kernels to what is effectively required. A first approach consists in adapting the degree of freedom of the output Student kernel based on the dimensionality with the following kernels [181, 183]:

$$
\beta_{ij} \triangleq \frac{\exp\left(-\frac{\Delta_{ij}^2}{2\sigma_i^2}\right)}{\sum_{k\neq i}\exp\left(-\frac{\Delta_{ik}^2}{2\sigma_i^2}\right)} \quad \text{and} \quad b'_{ij} \triangleq \frac{\left(1+\frac{D_{ij}^2}{l}\right)^{-\frac{l+1}{2}}}{\sum_k \sum_{m\neq k}\left(1+\frac{D_{km}^2}{l}\right)^{-\frac{l+1}{2}}}, \quad (5.22)
$$

with $l \triangleq d - 1$. If the embedding space dimensionality d is chosen very high, this tends towards SNE. Yet, this does not take into account the effective dimensionality of the data.

The degree of freedom l in Eq. (5.22) may also be learnt from the data, as in Heavy-tailed Symmetric SNE (HSSNE), using it as an optimization variable for the stress function [181, 202]. This allows every values of $l \in \mathbb{R}^*$, and in the absence of crowding problem, l should tend towards $+\infty$. However, in that approach, there is a strong interdependence between the embedded points coordinates and the degree of freedom l defining the stress function. From an optimization stand-point, this definition is thus not quite satisfactory.

Considering that the intrinsic dimensionality is a local property, Inhomogeneous tSNE [97] adapts the previous approach by defining point-wise values the degree of freedom l_i of the kernel in the embedding space. In that case, the membership degrees are not symmetrized:

$$
\beta_{ij} \triangleq \frac{\exp(-\Delta_{ij}^2/2\sigma_i^2)}{\sum_{k\neq i}\exp(-\Delta_{ik}^2/2\sigma_i^2)} \quad \text{and} \quad b_{ij} \triangleq \frac{\left(1+\frac{D_{ij}^2}{l_i}\right)^{-\frac{l_i+1}{2}}}{\sum_{k\neq i}\left(1+\frac{D_{ik}^2}{l_i}\right)^{-\frac{l_i+1}{2}}}.
$$

The dimensionality of data may also be accounted for by setting the data space kernel based on an estimation of the intrinsic dimensionality. Intrinsic tSNE (itSNE) [161] uses a power kernel to define β_{ij}, leading to even thinner tail than the Gaussian (see Fig. 5.5b):

$$
\beta_{ij} \triangleq \frac{\exp\left(-\frac{\Delta_{ij}^{\lambda_i}}{2\sigma_i^2}\right)}{\sum_{j\neq i}\exp\left(-\frac{\Delta_{ik}^{\lambda_i}}{2\sigma_i^2}\right)} \quad \text{and} \quad b'_{ij} \triangleq \frac{(1+D_{ij}^2)^{-1}}{\sum_k \sum_{l\neq k}(1+D_{kl}^2)^{-1}},
$$

with $\lambda_i = \partial_i$. The intrinsic dimensionality ∂_i is estimated locally by the Hill estimator [6]. We may note that the embedding space kernel of itSNE is not adapted to the power kernel family, keeping the Student kernel with one degree of freedom. The method's purpose being the representation of outliers, the excessive stretching that ensues is not considered problematic.

The twice Student Stochastic Neighbour Embedding (ttSNE) [47] uses Student kernels in both spaces. The degrees of freedom of those are set based on the intrinsic dimensionality in one space, and the embedding dimensionality in the other, leading to:

$$\beta_{ij} \triangleq \frac{\left(1 + \frac{\Delta_{ij}^2}{\lambda \sigma_i^2}\right)^{-\frac{\lambda+1}{2}}}{\sum_{k \neq i}\left(1 + \frac{\Delta_{ik}^2}{\lambda \sigma_i^2}\right)^{-\frac{\lambda+1}{2}}} \quad \text{and} \quad b'_{ij} \triangleq \frac{\left(1 + \frac{D_{ij}^2}{l}\right)^{-\frac{l+1}{2}}}{\sum_k \sum_{m \neq k}\left(1 + \frac{D_{km}^2}{l}\right)^{-\frac{l+1}{2}}},$$

(5.23)

with $\lambda \triangleq \partial - 1$ and $l \triangleq d - 1$. The intrinsic dimensionality $\widehat{\partial}$ is estimated globally with the method proposed by Lee et al. [110].

5.3.7 Adaptive Student Kernel Imbedding

We introduce here the Adaptive Student Kernel Imbedding (ASKI). This neighbourhood embedding technique relies on Student kernels whose degrees of freedom automatically adapt to account for the effective dimensionality discrepancy between the data and embedding spaces. Indeed, the membership degrees are given by:

$$\beta_{ij} \triangleq \frac{\left(1 + \frac{\Delta_{ij}^2}{\lambda_i \sigma_i^2}\right)^{-\frac{\lambda_i+1}{2}}}{\sum_{k \neq i}\left(1 + \frac{\Delta_{ik}^2}{\lambda_i \sigma_i^2}\right)^{-\frac{\lambda_i+1}{2}}} \quad \text{and} \quad b_{ij} \triangleq \frac{\left(1 + \frac{D_{ij}^2}{l s_i^2}\right)^{-\frac{l+1}{2}}}{\sum_{k \neq i}\left(1 + \frac{D_{ik}^2}{l s_i^2}\right)^{-\frac{l+1}{2}}}.$$

The degrees of freedom are set as $\lambda_i \triangleq \max(\partial_i, d)$ and $l \triangleq d$, where ∂_i is the local intrinsic dimensionality at point i. Thus, if the embedding dimensionality is to small compared with the manifold intrinsic dimensionality, the points are allowed to spread out more in the map.

The data space scale σ_i is set by a fixed perplexity $\kappa^\star$, and the embedding space scale s_i is set either as $s_i = \sigma_i$ or as $s_i = 1$ (metric or non-metric version). Note that for a given scale σ_i the perplexity increases when the degree of freedom λ_i of the kernel decreases. In addition, for small values of λ_i, the perplexity can be lower-bounded by a value much higher than 1, so that some neighbourhoods may not reach the desired value.

The estimation of the intrinsic dimensionality ∂_i is a complicated problem with no clear solution (see Sect. 2.2). Here, we use the covariance dimensionality [45] assessed on a $\kappa^\star$-neighbourhood with $\kappa^\star$ the perplexity. The dimensionality is set to account for exactly 95% of the variance, allowing non-integer values. This is obtained by linear interpolation of the number of components as a function of the explained variance (see Sect. 2.2.1.2).

Those kernels are then compared with a mixture of Kullback–Leibler divergence:

$$\zeta_i^{\text{ASKI}} \triangleq \mathcal{M}_{\text{KL},12}(\beta_i, b_i, \tau).$$

This stress is optimized with the Broyden–Fletcher–Goldfarb–Shanno (BFGS) quasi-Newton algorithm [138], using cMDS initialization [177] and a multi-scale optimization approach [110]. Those different elements are detailed in Sect. 7.1. Since the intrinsic dimensionality estimate depends on the scale $\kappa^\star$ it is re-computed for each value of the perplexity.

As opposed to Inhomogeneous tSNE, the degree of freedom of ASKI is not considered as a free parameter obtained by optimizing the stress for a given map. Thereby, the coordinates of the embedded points are the only elements affecting the stress. This ensures that the stress function does not depend on the position of points, which we consider to be a desirable feature for a stress function.

To show the specific impact of the kernel adaptation we focus on the metric case with a trade-off $\tau = 1$ and perplexity $\kappa^\star = 32$, comparing ASKI with alternate versions of SNE and tSNE (metric and non-symmetrized). This amounts to consider degrees of freedom $\lambda = \infty$, and $l = \infty$ or $l = 1$. This comparison is performed for three datasets:two open boxes, digits (with 512 randomly selected samples) and oil flow introduced Sect. 1.1.7). The obtained maps are presented Figs. 5.6, 5.7 and 5.8, along with their associated Shepard diagrams, accounting for the preservation of distances. The rank-based quality is shown Fig. 5.9 using the average of trustworthiness and continuity indicators $(\check{\mathcal{T}} + \check{\mathcal{C}})/2$.

We first consider the two open boxes, which constitute a dataset of rather low dimensionality. In the maps shown Fig. 5.6a–c, all three methods manage to represent the boxes by unfolding them in the two-dimensional space. However, tSNE differs from the others by inducing a very important distance between the boxes. This is confirmed by the Shepard diagrams (Fig. 5.6d–f). For SNE and ASKI the Shepard are both rahter close to a straight line, whereas for tSNE there is a strong gap between high and small distances. Thus, in this low dimensional case, tSNE unnecessarily stretches the high distances compared to the small distances, although the true distances could be better preserved due to the low discrepancy between the manifold and embedding space dimensionality. As for the preservation of neighbourhood ranks (Fig. 5.9a), the performances of all methods are almost equivalent.

For the very high-dimensional digits dataset, ASKI is closer from tSNE than from SNE. The maps (Fig. 5.7a–c) show better separation of the digits classes for ASKI and tSNE, which seems to correspond to a better preservation of the

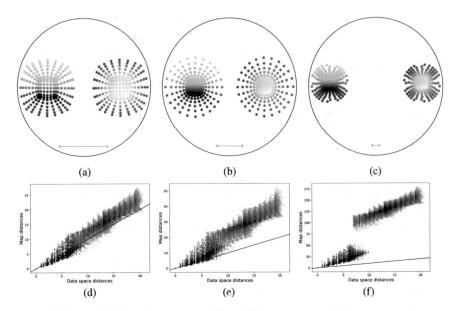

Fig. 5.6 ASKI results for the embedding of the two open boxes dataset of low dimensionality. Colours are RGB encoding of the data space coordinates of each point. Scales added on the maps indicate the data radius $\Delta_{max}/2$. Shepard diagrams show the joined distribution of data space and map distances. tSNE appear to over separate the two boxes. (**a**) SNE map. (**b**) ASKI map. (**c**) tSNE map. (**d**) SNE Shepard. (**e**) ASKI Shepard. (**f**) tSNE Shepard

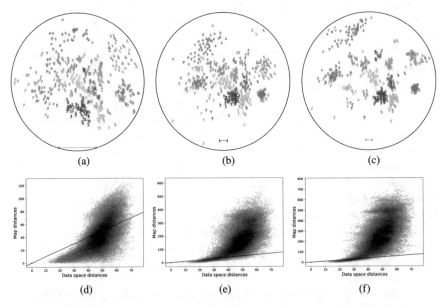

Fig. 5.7 ASKI results for the digits dataset of high dimensionality. (**a**) SNE map. (**b**) ASKI map. (**c**) tSNE map. (**d**) SNE Shepard. (**e**) ASKI Shepard. (**f**) tSNE Shepard

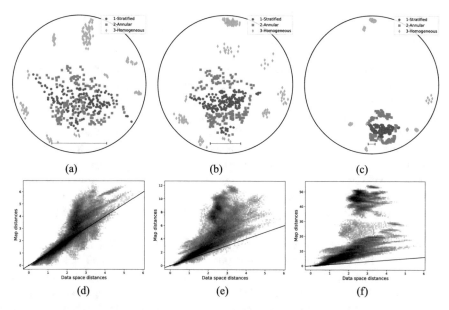

Fig. 5.8 ASKI results for the oil flow dataset of intermediate dimensionality. (**a**) SNE map. (**b**) ASKI map. (**c**) tSNE map. (**d**) SNE Shepard. (**e**) ASKI Shepard. (**f**) tSNE Shepard

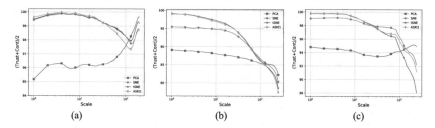

Fig. 5.9 $(\check{\mathcal{T}} + \check{\mathcal{C}})/2$ as a function of the scale κ for maps of the three datasets shown Fig. 5.6, 5.7 and 5.8. Quality evaluation of the PCA map is also provided for reference. (**a**) Two open boxes. (**b**) Digits. (**c**) Oil flow

data structure as shown Fig. 5.9a. This may be analysed in terms of distances preservation (Fig. 5.7d–f), seeing that ASKI and tSNE allow important stretching of the high distances. Hence, those methods allow to convert the excessive dimensional complexity into spatial spreading. Conversely, SNE, which leads to a far better preservation of distances, has to distort the neighbourhood structure. Hence, in this case the stretching of high distances by ASKI and tSNE may be deemed necessary for the preservation of the local neighbourhood structure.

Finally, we focus on the oil flow dataset of intermediate dimensionality (Figs. 5.8 and 5.9c). In that case, tSNE represents clusters of the third class as excessively distant from the first and second classes, as confirmed by the gap in the Shepard diagram. Inversely, SNE mixes the first and second classes, leading to slightly worst performances in terms of quality evaluation for the small scales κ. ASKI provides an intermediate solution, preserving better the small neighbourhoods, while avoiding to excessively stretch the distances.

In coherence with their kernel structures, SNE is better suited for embedding rather low-dimensional data, and tSNE for very high dimensional data, while ASKI benefits from its adaptive kernels. Thereby, this technique provides an automatically managed trade-off between the two others. It preserves the rank-based neighbourhood structure as well as the best method, but also the distances if possible, that is, when the dimensionality discrepancy is low. In addition, it outperforms both methods for intermediate dimensionality discrepancy. Finally, avoiding excessive gaps leads to a better trade-off between the representation of clusters separation and of the internal structure of each of those structure. Indeed, with a fixed level of zoom, a map with huge gaps between clusters provides far less details for each individual cluster.

We may note that the use of heavy-tailed kernels in ASKI leads to several limitations. First, it is sometimes not possible to reach very low values of the perplexity numerically. In the case of Gaussian kernels in the data space, a perplexity of one, may be reached, as long as the distance to the nearest neighbour is strictly lower than the distance to the second nearest neighbour. Indeed, when $\sigma_i \to 0$, the exponential term associated to the smallest distance becomes predominant compared with the one associated to the second smallest distance, as well as all other exponential terms. Hence, the membership degree associated to the nearest neighbour is equal to one, while all the others are zero. Yet, the Student kernel does not tend to zero as fast as the Gaussian kernel when $\sigma_i \to 0$, especially for low values of λ_i. Hence, for computationally tractable values of σ_i, the perplexity may remain higher than expected. A second limitation associated to heavy-tails is that of mixturability. Indeed, we have noticed that the variability of maps with the trade-off parameter is lower with Student kernels of low degrees of freedom than with Gaussian kernels. This may come from the fact that, since the terms are never really close to zero with heavy-tailed kernels, soft recall and soft precision provide a less different penalization of missed and false neighbours than with Gaussian kernels.

Future work should include detailed comparison of ASKI with the closely defined ttSNE. In terms of definition the main differences between these two methods are the following: the degree of freedom in ASKI is set locally, their is no symmetrization step, and ASKI incorporate a metric behaviour as well as possible mixturation of its divergence. One of the differences that should appear is that of outlier representation, since symmetrization tend to erase outliers.

Work could also be conducted to show the effect of adaptability for embedding data in higher dimensional spaces. Other local intrinsic dimensionality estimation methods could be employed for ASKI, such as TIDLE. The choice of the kernel family could also be challenged, for example using power kernels (see Fig. 5.5b), which could reduce the limitations noticed with heavy-tailed kernels.

5.4 Graph Layout

Graph-layout techniques aim at embedding network data in a feature space by mapping each vertex to an embedded point. This embedding should represent close the pairs of points connected by an edge, especially if its weight is high, and apart the points that are not connected by an edge. Graph layout is strongly linked with dimensionality reduction since metric data may be converted into network data, for instance using a proximity graph (e.g., κ-Nearest Neighbour graph). Therefore, a similarity and/or dissimilarity matrix may be interpreted as the adjacency matrix of a weighted complete graph.

Graph layout techniques, such as Elastic Embedding (EE) [31], LargeVis [174] and Uniform Manifold Approximation and Projection (UMAP) [125] may be related to Neighbourhood Embedding [20, 204], since they share a common goal of preserving neighbourhood relations. From a technical perspective, both may be perceived as comparing based on a divergence measure the pairwise similarities in the data and embedding space. Yet, they differ by the separability of their divergence measure [204]. Indeed, graph-layout methods focus on each edge independently, allowing to separate the stress ζ into a sum of pairwise sub-stresses ζ_{ij} each depending only on the associated distance D_{ij}. Conversely, Neighbourhood Embedding methods normalize the membership degrees, considering the distributions of neighbours around each point. This leads to interdependencies between all pairwise relations, the increase of proximity of one neighbour decreasing the membership degree of all the others. As such, graph-layout techniques leads to easier computations.

5.4.1 Force Directed Graph Layout: Elastic Embedding

The Elastic Embedding (EE) [31] uses as force-directed layout approach of optimization detailed more in Sect. 7.1.5. As such, it decomposes its stress between two components respectively penalizing the remoteness and proximity between pairs of points. Hence, the gradient descent minimization of those sub-stresses correspond to exerting attractive or repulsive forces between the points. The forces being weighted depending on the effective similarity or dissimilarity of points in the data space, this minimization leads to an equilibrium state where the position of points reflect the data structure. That stress is expressed by:

$$\zeta_i^{\mathrm{EE}} \triangleq \sum_{j \neq i} w_{ij}^{+} D_{ij}^2 + s^2 \sum_{j \neq i} w_{ij}^{-} \exp(-D_{ij}^2),$$

with s a scaling parameter, w_{ij}^{+} and w_{ij}^{-} attractive and repulsive weights for the pair (i, j) given by:

$$w_{ij}^{+} \triangleq \exp\left(-\frac{\Delta_{ij}^2}{2\sigma^2}\right) \quad \text{and} \quad w_{ij}^{-} \triangleq \Delta_{ij}^2.$$

w_{ij}^+ and w_{ij}^- could be seen here as weights of two complete graphs (one with positive edges and the other with negative edges).

5.4.2 Probabilistic Graph Layout: LargeVis

LargeVis [174] performs a layout of the κ-nearest neighbours graph in the data space denoted here $(V, E(\kappa))$, based on the a probabilistic model. Edges $(i, j) \in E(\kappa)$ existing in that graph are referred to as *positive edges*, while the others $(i, j) \notin E(\kappa)$ are coined *negative edges*. The weights γ_{ij} of that graph are defined similarly to the symmetrized data space membership degrees β'_{ij} of tSNE (with Gaussian kernels and perplexity setting), but restricting the normalization the pairs $(i, j) \in E(\kappa)$:

$$\gamma'_{ij} \triangleq \frac{\gamma_{ij} + \gamma_{ji}}{2N} \text{ with } \gamma_{ij} \triangleq \frac{\exp(-\Delta_{ij}^2/2\sigma_i^2)}{\sum_{(i,k)\in E(\kappa)} \exp(-\Delta_{ik}^2/2\sigma_i^2)}.$$

The probabilistic model attempts to assess the likelihood of a given layout with vertices embedded at positions $\{x_i\}$. For that purpose, the probability g_{ij} that an edge with weight 1 exists between embedded vertices i and j, knowing a given layout, is modelled by a decreasing function of the distance D_{ij} between these vertices. This gives:

$$g_{ij} \triangleq \frac{1}{1 + D_{ij}^2}.$$

The likelihood of a positive edge (i, j) with weight γ'_{ij} is then modelled as $g_{ij}^{\gamma'_{ij}}$ (i.e. for the same distance, the likelihood is higher for an edge with small weight γ'_{ij}). Conversely, the likelihood of the absence of a negative edge (i, j) is given by $(1 - g_{ij})^{\gamma^-}$, with γ^- an uniform weight for negative edges. The stress of LargeVis is then obtained as the negative log-likelihood (to be minimized) of the graph constituted of all those edges, which is:

$$\zeta_i \triangleq - \sum_{(i,j)\in E(\kappa)} \gamma'_{ij} \log g_{ij} - \sum_{(i,j)\notin E(\kappa)} \gamma^- \log(1 - g_{ij}).$$

5.4.3 Topological Method UMAP

Uniform Manifold Approximation and Projection (UMAP) [125] provides an interpretation of its probabilistic graph layout based on fuzzy topology. Indeed, it assumes that the underlying data manifold may be approximated by a simplicial complex (e.g., a set of simplices), which may be summarized, for the sake of tractability, by its 1-skeleton (that is the set of edges of those simplices). This set is considered fuzzy, meaning that the belonging of edges to that set is assessed by a membership strength between 0 and 1 (decreasing with the edge length), rather than a Boolean value. Thereby, UMAP describes the structures of the data manifold and embedding by two fuzzy graphs, which are then compared by a divergence measure to be minimized.

The graph in the data space is defined as the κ-nearest neighbours graph. Its edges (i, j) have a membership strength γ_i given by:

$$\gamma_{ij} \triangleq \exp\left(-\frac{(\Delta_{ij} - \min_{k \neq i} \Delta_{ik})^+}{\sigma_i}\right).$$

The definition of γ_{ij} (subtracting the minimum of distances from i) replicates the property of shift-invariance of SNE-like methods, without the normalization. The scale parameter σ_i is set so that the cardinality of the fuzzy set of edges starting from i reaches a fixed value determined empirically:

$$\sum_{j \in v_i(\kappa)} \gamma_{ij} = \log_2(\kappa).$$

A symmetrization is then performed using a fuzzy set union:

$$\gamma'_{ij} \triangleq \gamma_{ij} + \gamma_{ji} - \gamma_{ij}\gamma_{ji},$$

or fuzzy set intersection:

$$\gamma'_{ij} \triangleq \gamma_{ij}\gamma_{ji}.$$

In the embedding, the membership strengths g_{ij} are defined by:

$$g_{ij} \triangleq \frac{1}{1 + (D_{ij}/s)^{2l}}.$$

The parameters s and l are set so that the above smooth expression best approximate the following:

$$g_{ij}^{\star} \triangleq \exp(-(D_{ij} - D_{\min})^+),$$

with $D_{\min}$ a user set parameter indicating the minimum expected distance between embedded points.

The membership are compared between the two spaces by the Kullback–Leibler divergence of the distribution $(\gamma'_{ij}, 1 - \gamma'_{ij})$ and $(g_{ij}, 1 - g_{ij})$ (membership and non-membership):

$$\zeta_i^{\text{UMAP}} \triangleq \sum_j \gamma'_{ij} \log\left(\frac{\gamma'_{ij}}{g_{ij}}\right) + (1 - \gamma'_{ij}) \log\left(\frac{1 - \gamma'_{ij}}{1 - g_{ij}}\right).$$

Thus it measures the inadequacy of the memberships but also of the non-membership (probability that the edge is not in the fuzzy simplicial set).

5.5 Artificial Neural Networks

Artificial neural networks are parametric models capable of approximating a wide variety of functions, as shown by several Universal Approximation Theorems [119].

5.5.1 Auto-Encoders

Auto-encoders [88] use artificial neural networks to encode the data in a lower dimensional form. These networks may be split into two parts: an encoder and a decoder (see Fig. 5.10). The encoder part encodes the high dimensional data points ξ_i by lower dimensional points x_i. This encoder $\Phi_{(W_1,...,W_L)} : \mathbb{R}^\delta \longrightarrow \mathbb{R}^d$ is constituted of L layers with a progressively decreasing number of neurons (also called units). The parameters of the encoder network are given by matrices $\{W_k\}$. For a point i, the first layer takes the data space coordinates ξ_i as input and the last layer (which is the code layer) returns the embedding space coordinates x_i as output. Conversely, the decoder part $\Psi_{(W_L^\top,...,W_1^\top)} : \mathbb{R}^d \longrightarrow \mathbb{R}^\delta$ takes the low-dimensional representations and searches to reconstruct the data points, by using successive layers with progressively increasing number of neurons. We may note that, in this simple fully connected architecture, the weights matrix for symmetric layers are the transpose of one another, in order to constrain the parameter space. The hidden layers use logistic units, namely neurons with logistic activation functions, and the code layer use linear units.

Similarly to PCA, auto-encoders constitute a parametric DR method, for which the parameters of a low rank model, namely the weights of the auto-encoder network, are chosen so as to minimize the reconstruction error for the dataset. In this case, the equivalent of the low rank constraint comes from the bottleneck of the central layer with d neurons. This reconstruction error may be expressed as the loss function:

$$\zeta \triangleq \left\| \Psi_{(W_L^\top,...,W_1^\top)} \circ \Phi_{(W_1,...,W_L)}(\Xi) - \Xi \right\|_2,$$

with $\| \bullet \|$ the Frobenius norm.

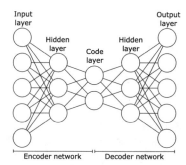

Fig. 5.10 Neural network architecture for autoencoders, with code layers forming a bottleneck with one unit by embedding dimensionality. This network is trained to minimize the reconstruction error between the input and output layers for the dataset. Here, the represented network encodes five-dimensional data in a two-dimensional space

In order to minimize the loss function with respect to the weights, gradient descent may not be used directly with random initialization of the weights, since it often fails to converge toward an acceptable local minimum. Thus, a pre-training phase using restricted Boltzmann machines is necessary to obtain a good initialization of the weights, which may be followed by gradient descent fine-tuning. The gradient of this stress with respect to all intermediary weight matrices may be computed efficiently using back-propagation.

Auto-encoders are a powerful tool for unsupervised dimensionality reduction due to the approximation property of neural networks. Furthermore, they may be far more suited for extracting features from data with high semantic levels, such as complex images or texts. This gives them a significant interest for use in a machine learning pipeline. Yet, they suffer from interpretability issues due to their complex and unintuitive structure. Thus, for human in the loop applications, other methods such as MDS and NE with the clearly stated goal of distance or neighbourhood preservation, may be easier to grasp from a human analyst perspective.

5.5.2 *IVIS*

IVIS [173] is a parametric method designed to preserve the global structure. It uses an encoder neural network trained by minimizing a triplet loss to ensure that the map preserves some randomly selected κ-neighbours and κ non-neighbours relations. For a given anchor point i, a neighbour j and a non-neighbour k are randomly selected respectively in the sets of κ nearest neighbours and κ non-neighbours (see Fig. 5.11). These may also be seen as a query i, a positive result j and a negative result k for that query, in an information retrieval framework close from the one developed in [189]. The loss function is meant to guarantee that the embedding space distance D_{ij} between the anchor i and its considered neighbour j is lower, by a margin m, than the distances D_{ik} and D_{jk} separating those two points from the

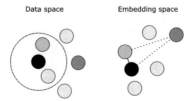

Data space   Embedding space

Fig. 5.11 Triplet loss of IVIS method. Considering the κ-neighbourhood (dashed circle) of the anchor (black point), a positive (green) and negative (grey) samples are selected. IVIS aim at maximizing distances between the anchor and positive samples (dotted line) compared with the distance between the anchor and positive sample (plain line)

selected non-neighbours k. That loss function for a given epoch is expressed by:

$$
\zeta \triangleq \left(\sum_{i,j,k\in\mathcal{T}} D_{ij} - \min(D_{ik}, D_{jk}) + m \right)^+ ,
$$

where $\mathcal{T}$ is the batch of triplets (i, j, k). These triplets are reselected at each epoch (i.e. at each full pass through the entire dataset), during the optimization of the stress. Hence, different neighbours j and non-neighbours k are successively considered for each point i. The use of a too high value for m is shown to give highly correlated embedding dimensions interpreted as a loss of information. The encoder network is designed as a Siamese neural network, simultaneously embedding the anchor, the positive result and the negative result. It is composed of three neural networks with similar weights using Scaled Exponential Linear Units (benefitting from a self-normalizing property [98]) for the hidden layers, and linear activations for the last layer.

5.5.3 Intermediate Conclusions

Many DR techniques have been developed in order to represent the structure of metric data in a low dimensional space. That structure may be expressed for instance in terms of distances (metric MDS), neighbourhood ranks (non-metric MDS), neighbourhood membership degrees (Neighbourhood Embedding), edges weights (graph layout), leading to different ways of interpreting the resulting embedding. In the case of Neighbourhood Embedding, techniques may benefit from the use of different kernels in the two spaces for assessing the membership degrees. Yet, as shown for ASKI, the difference between these kernels should account for the effective dimensionality discrepancy between the manifold and embedding space, and should not be fixed a priori. The different notions of data structure introduced for unsupervised DR may also be combined with side information (annotations), thus leading to the supervised DR methods detailed in the next chapter.

Chapter 6
Stress Functions for Supervised Dimensionality Reduction

One of the properties inherent in mathematics is that any real progress is accompanied by the discovery and development of new methods and simplifications of previous procedures ... The unified character of mathematics lies in its very nature; indeed, mathematics is the foundation of all exact natural sciences.

David Hilbert

In the general case,t Dimensionality Reduction (DR) is an unsupervised task. Indeed, it does not necessitate data annotations, as opposed to classification for which the desired output must be provided for a training set. Yet, DR is sometimes applied to datasets for which such annotations are available (e.g., class-information). In this context, supervised dimensionality reduction methods seek to take advantage of that information to improve the representation of the data structure. Supervised dimensionality reduction appears valuable in two main contexts [75, 79, 116]:

- Pre-processing of data for further machine learning applications, such as classification: in this case, the purpose is often to emphasize class separation in the embedding space, possibly distorting the data structure.
- Visual exploration of labelled data: in that case we consider that the main goal is to preserve the structure of data, as for unsupervised dimensionality reduction. This is combined with a secondary goal of limiting the distortions of the class structure.

This Chapter first presents the different types of class-information that may be available (Sect. 6.1). Then, several families of supervised DR are detailed distinguishing different approaches. Section 6.2 describes parametric methods maximizing class separation in the map. Then, Sect. 6.3 presents methods combining unsupervised DR and metric learning to emphasize class separation. Finally, methods tuning the parameters governing the preservation of the structure are detailed. Section 6.4 focuses on the tuning of the neighbourhood size based on class structure, while Sects. 6.5 and 6.6 consider differentiated penalization of false and missed neighbours for intraclass and interclass relations in order to minimize the distortions mostly affecting the class structure.

6.1   Types of Supervision

Annotations vary across datasets, leading to different types of supervision. Those may be described as fully supervised , weakly supervised or semi-supervised, relying on the terminology detailed by Bellet et al. [15]. Those annotations are illustrated Fig. 6.1.

6.1.1   Full Supervision

In the fully supervised case (illustrated in Figs. 6.1a–c), each point ξ_i is associated with a label indicating to which class(es) it belongs, those classes constituting n clearly defined categories. In the simplest case, those labels are *hard labels* of the form $L_i \triangleq k$, unambiguously assigning the point i to the kth of the n classes. In this framework, we denote $\{C_k\}$ the sets of points indices belonging to these classes defined as $C_k \triangleq \{i \in [\![1; N]\!] \mid L_i = k\}$. In the case of human annotated data, labels are most often hard labels. Labels may also be *soft labels* quantifying the uncertainty of a point belonging to one class or another. A soft label $\tilde{L}_i \triangleq (\tilde{L}_{ik})_{k \in [\![1;n]\!]}$ is a row vector providing a discrete probability distribution over all classes with the element $\tilde{L}_{ik}$ indicating the probability that the point ξ_i belongs to the kth class. Soft labels may for instance be obtained using a classifier or a clustering algorithm. To

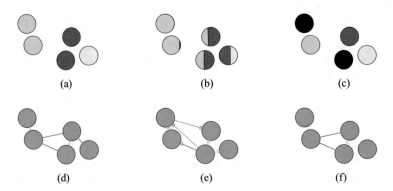

(a)   (b)   (c)

(d)   (e)   (f)

Fig. 6.1 Different types of supervisions. In the case of full supervision labels are presented in colours dividing points between the cyan, red and yellow classes. For soft labels, areas of each colour indicate the probability to belong to each class. Points with missing labels are shown in black. For pairwise information, pairs of points can be labelled as similar (green links) or as dissimilar (magenta links). In the triplet case, an anchor (or query) points towards a positive example (green arrow) and a negative example (magenta arrow), the positive being more similar to the anchor than the negative. An anchor may be associated with several pairs of positive/negative examples. (**a**) Hard labels (full supervision). (**b**) Soft labels (full supervision). (**c**) Hard labels (full semi-supervision). (**d**) Pairwise annotations (weak supervision). (**e**) Triplet annotations (weak supervision). (**f**) Pairwise annotations (weak semi-supervision)

combine the two frameworks, hard labels can be treated as soft labels by converting them to one-hot encoded vectors (i.e. vectors for which one of the coordinate is one and the others are zeros). Note that many other cases may be considered. For instance, multi-labels can indicate the simultaneous belonging of a point to several distinct classes. Furthermore, classes can be organized in a hierarchical structure, with several classes being regrouped in a more general classes. In the case of diagnosis, multi-labels can be used to indicate that a system is simultaneously subject to several kinds of faults, while hierarchical class structures may regroup several highly specific faults into a broader category.

6.1.2 Weak Supervision

In the weakly supervised case (illustrated in Figs. 6.1d–f), classes are not always defined explicitly. Thus, information is not given for each point individually, but rather with respect to the others. Weak annotations are mostly provided for pairs or triplets of points [15].

Pairwise annotations determine a set $S^{\in}$ of pairs of points (i, j) that are semantically similar, as well as a set $S^{\neq}$ of point pairs that are semantically dissimilar. As such, they identify "must be" and "must not be" pairwise relations. Full supervision naturally induces weak supervision by defining:

$$S^{\in} \triangleq \{(i, j) \mid L_i = L_j\} \quad \text{and} \quad S^{\neq} \triangleq \{(i, j) \mid L_i \neq L_j\}.$$

Yet, in the general case, the pairwise relations established between data points are not necessarily transitive. Hence, it may not be possible to extract equivalence classes from pairwise annotations.

Weak supervision may also provide information for triplets (i, j, k), specifying that point ξ_i is semantically more similar to point ξ_j than to ξ_k. It is used in the case of queries for information retrieval, identifying a query ξ_i, a positive result ξ_j along with a negative result ξ_k.

6.1.3 Semi-Supervision

Both full and weak supervisions may be semi-supervised if only part of the data is subject to supervision, as shown in Figs. 6.1c and f. It means that certain points have missing labels (denoted here by $L_i = -1$), or are not involved in any pair or triplet information. The set of pairs that are not semantically characterized is denoted by $S^{\emptyset}$.

6.2 Parametric with Class Purity

Some supervised Dimensionality Reduction methods use a parametric transformation maximizing the class purity in the embedding space.

6.2.1 Linear Discriminant Analysis

Linear Discriminant Analysis (LDA) [67], sometimes called Fisher Discriminant Analysis (FDA), performs an orthogonal linear projection defined by a matrix P (of size $\delta \times d$). This projection is the one maximizing the ratio between the interclass and intra-class variances in the embedding space. It is designed to optimally separate classes each constituted of one Gaussian cluster, with the assumption of homoscedasticity (i.e. the common covariance for all classes). In that regard, when applying LDA, the embedding dimensionality d is usually set to $d \triangleq n - 1$ (with n being the number of classes). Formally, it solves for each column p_j of the matrix P (constrained to be orthogonal to all the others) the following optimization problem:

$$\underset{p^\top p=1}{\operatorname{argmax}} \left(\frac{p^\top C^{\not\in}_{\Xi} p}{p^\top C^{\in}_{\Xi} p} \right).$$

The covariance matrix $C^{\in}_{\Xi} \triangleq \frac{1}{N} \tilde{\Xi}^\top_C \tilde{\Xi}_C$ shows the spreading within the classes (i.e. the variance of each class), with $\tilde{\Xi}_C$ the $N \times \delta$ matrix of points centred by their class centroids, namely with rows $\xi_i - \mu_{L_i}$. Conversely, the covariance matrix $C^{\not\in}_{\Xi} \triangleq \tilde{M}^\top_C \tilde{M}_C$, measures the the spreading between the classes (spread of class centroids), with $\tilde{M}_C$ being the $n \times \delta$ matrix of centred class centroids, with rows $\mu_k - \bar{\mu}$. The aforementioned class centroids are defined as $\mu_k \triangleq \frac{1}{|C_k|} \sum_{i \in C_k} \xi_i$ and their centroid is $\bar{\mu} \triangleq \frac{1}{n} \sum_k \mu_k$. This problem is solved by eigenvalue decomposition of $(C^{\in}_{\Xi})^{-1} C^{\not\in}_{\Xi}$.

As a historic method for supervised dimensionality reduction, LDA has many variants. Kernel Discriminant Analysis (KDA) [128] uses the kernel trick to perform LDA from a basis expansion space in a computationally tractable way. Fisher Locally Linear Embedding (FLLE) [51] combines LDA with LLE with a trade-off parameter balancing the supervised and unsupervised objectives of these two methods. Local Fisher Discriminant Analysis (LFDA) [172] adapts LDA to manage the case of multi-modal classes.

6.2.2 Neighbourhood Component Analysis

Neighbourhood Component Analysis (NCA) [79] finds a linear transformation, defined by a matrix M, maximizing the class purity of a neighbourhood. As such, it

is designed to maximize classification performances in the map for a leave-one-out
k-NN classifier. The neighbourhood structure in the embedding space is assessed
using the neighbourhood membership degrees $\{b_{ij}\}$, defined as for Neighbourhood
Embedding methods:

$$b_{ij} \triangleq \frac{\exp(-||\xi_i M - \xi_j M||_2^2)}{\sum_{k \neq i} \exp(-||\xi_i M - \xi_k M||_2^2)}. \tag{6.1}$$

The sum $\sum_{j \in C_k} b_{ij}$ models the proportion of point i embedding space neighbours
that belong to class k. Hence, maximizing the class purity (proportion of intraclass
neighbours in the neighbourhood) amounts to minimize one of the smooth stresses:

$$\zeta_i^{NCA} \triangleq - \sum_{j \in C_{L_i}} b_{ij} \quad \text{or} \quad \zeta_i^{NCA'} \triangleq - \log \left(\sum_{j \in C_{L_i}} b_{ij} \right).$$

This optimization problem is also equivalent to minimizing the error of a k-NN
classifier returning soft labels. Indeed, the error between soft labels $\{\tilde{L}_i\}$ (probability
distribution over the classes) and the distribution of classes in the embedding space
neighbourhoods is the following (relying either on the L_1 norm or the Kullback–
Leibler divergence):

$$\zeta_i^{NCA} \triangleq \sum_k \left| \tilde{L}_{ik} - \sum_{j \in C_k} b_{ij} \right| \quad \text{or} \quad \zeta_i^{NCA'} \triangleq \sum_k \tilde{L}_{ik} \log \left(\frac{\tilde{L}_{ik}}{\sum_{j \in C_k} b_{ij}} \right). \tag{6.2}$$

Considering that the $\{\tilde{L}_i\}$ are hard labels (with one-hot encoded vector), the soft
labels formulation of Eq. (6.2) leads back to Eq. (6.1).

An alternative version of NCA uses another parametric model, replacing the linear transform by an encoder neural network (pre-trained with Restricted Boltzmann
Machines and fine-tuned with NCA stress) [157].

6.3 Metric Learning

Another approach for supervised Dimensionality Reduction combines non-
parametric unsupervised DR methods with metric learning. This is the case, for
example, of S-SNLM [144], S-Isomap [75], S-LLE and its variants [50, 207, 208],
S-NeRV [145] and S-UMAP [125]. Metric learning consists in defining a
data-dependent measure of distances often taking into account the supervision
information. It is also used with classification or clustering algorithms taking
distances as input (e.g., k-NN and k-means).

6.3.1 Mahalanobis Distances

A well-known case of metric adapting to the data is the generalized Mahalanobis distance [15], defined for a given matrix M of size $\delta \times r$, with r the rank of the metric. It is the distance associated with the norm defined by the quadratic form based on the symmetric positive semi-definite matrix $MM^\top$, that is:

$$\Delta_M(\xi_i, \xi_j)^2 \triangleq \sqrt{(\xi_i - \xi_j)MM^\top(\xi_i - \xi_j)^\top}.$$

For the classical Mahalanobis distance, M is defined based on the inverse of the covariance matrix, that is $M^\top M = C_\Xi^{-1}$.

The generalized Mahalanobis distance may be interpreted as the Euclidean distance on a transformed feature space obtained from the original data space by a linear transformation represented by the matrix M:

$$\Delta_M(\xi_i, \xi_j)^2 = \|\xi_i M - \xi_j M\|_2^2.$$

In that regard, methods such as NCA may be applied to learn a metric (i.e. define a matrix M) taking class-information into account (see Fig. 6.2). The dimensionality of the transformed space corresponds to the rank of the metric. Thus, we can see that metric learning and dimensionality reduction are strongly linked, with dimensionality reduction appearing as a rank-constrained metric learning, while supervised metric learning consitutes a pre-processing tool for dimensionality reduction methods using distances as input.

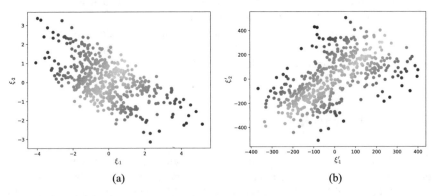

(a)   (b)

Fig. 6.2 Transformation of data with the Mahalanobis metric obtained with the full rank NCA [79]. The left panel presents the original data, while the right panel shows the transformed data, such that the learnt metric is the standard Euclidean distance in that transformed space. (**a**) Original data. (**b**) Data in transformed space

6.3.2 Riemannian Metric

Conversely to generalized Mahalanobis distances, defined globally in the ambient space, Riemannian metrics [144] are defined locally along the manifold as:

$$\Delta(\xi, \xi + d\xi) \triangleq \sqrt{d\xi \, M(\xi) M(\xi)^\top d\xi^\top},$$

where $d\xi \to 0$. The matrix $M(\xi)$ depends on the position. The proximity between ξ and $\xi + d\xi$ ensures that $M(\xi + d\xi) \approx M(\xi)$, thus preserving the symmetry property for the metric. This definition is then extended globally by integrating along the shortest path along the manifold:

$$\Delta(\xi_i, \xi_j) \triangleq \inf_{\substack{\xi:[0,1] \longrightarrow M \\ \xi(0) \triangleq \xi_i, \xi(1) = \xi_j}} \int_{t=0}^{1} \Delta(\xi(t), \xi(t + dt)).$$

This may be determined practically by the following steps. First, an estimation of the distance between each pair of points along a linear path is obtained by piecewise approximation along the segment (i, j) divided into T sub-segments:

$$\widehat{\Delta}_0(\xi_i, \xi_j) \triangleq \sum_{k=1}^{T-1} \Delta\left(\xi_i + \frac{k}{T}(\xi_i - \xi_j), \xi_i + \frac{k+1}{T}(\xi_i - \xi_j)\right).$$

This approximation along a linear path allows to account for the locality of the distance definition. Then, the shortest path distance between any two points is computed in the complete graph weighted by the piecewise approximations. Such a Riemannian metric is effectively used for the supervision of SNLM [144], NeRV [145] and tSNE [78], with $M(\xi)M(\xi)^\top$ defined as the Fisher information matrix.

6.3.3 Direct Distances Transformation

Other approaches directly modify the values of distances based on class-information. They use a different transformation for intra-class and inter-class distances yielding the following learnt distances:

$$\Delta'_{ij} \triangleq \begin{cases} f^{\in}(\Delta_{ij}), & \text{if } L_i = L_j \text{ (or } (i, j) \in \mathcal{S}^{\in}), \\ f^{\notin}(\Delta_{ij}), & \text{if } L_i \neq L_j \text{ (or } (i, j) \in \mathcal{S}^{\notin}) \end{cases}.$$

For a given value of Δ_{ij}, $f^{\notin}(\Delta_{ij}) > f^{\in}(\Delta_{ij})$ in order to amplify inter-class distances compared with intra-class distances. Then, unsupervised DR methods are directly applied to the learnt distances Δ'_{ij} instead of original distances Δ_{ij}.

6.3.3.1   Additive Transformation

Supervised Locally Linear Embedding (S-LLE) [50] adds a constant to the inter-class distances, leading to the following transformations of distances:

$$f^{\in}(\Delta_{ij}) = \Delta_{ij} \quad \text{and} \quad f^{\neq}(\Delta_{ij}) = \Delta_{ij} + \epsilon\Delta_{\max},$$

with $\Delta_{\max}$ being the data diameter (i.e. maximum of Δ_{ij} for all pairs (i, j)) and ϵ the level of supervision parameter in $[0, 1]$. Using the data diameter as additive value ensures that, for the maximum level of supervision $\epsilon = 1$, all learnt interclass distances will be higher than all intraclass distances.

This approach is extended to soft labels $\{\tilde{L}_i\}$ by the Probability-based Locally Linear Embedding (PLLE) [208]. The transformation of distances used by PLLE may be written as:

$$\Delta'_{ij} \triangleq \Delta_{ij} + (1 - q_{ij})\epsilon\Delta_{\max},$$

with $q_{ij} \triangleq \tilde{L}_i \cdot \tilde{L}_j$ a measure of class-community. In the case of hard labels, this class-community becomes $q_{ij} = 1_{L_i=L_j}$, the indicator function equal to 1 if i and j are of the same class and 0 otherwise. Thus, it leads back to S-LLE.

6.3.3.2   Multiplicative Transformation

Inversely, WeightedIso [192] modifies the intraclass distances through multiplication by a constant lower than 1. These learnt distances are then used to approximate geodesic distances, and perform the unsupervised Isomap algorithm. We may write this transformation using the level of supervision $\epsilon \in [0, 1]$ as

$$f^{\in}(\Delta_{ij}) = (1 - \epsilon)\Delta_{ij} \quad \text{and} \quad f^{\neq}(\Delta_{ij}) = \Delta_{ij}.$$

6.3.3.3   Concave vs Convex Transformations

S-Isomap [75] and Enhanced Supervised Locally Linear Embedding (ES-LLE) [207] respectively adapt Isomap and LLE using the same distance transformation. For this transformation, $f^{\in}$ is concave and $f^{\neq}$ is convex, allowing to amplify the effect of the metric learning for large distances. It is expressed as follows:

$$f^{\in}(\Delta_{ij}) = \sqrt{1 - \exp\left(-\frac{\Delta_{ij}^2}{\sigma}\right)} \quad \text{and} \quad f^{\neq} = \sqrt{\exp\left(\frac{\Delta_{ij}^2}{\sigma}\right)} - \Delta_0.$$

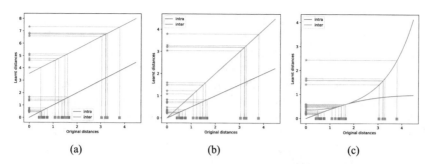

Fig. 6.3 Simple distance transformations for metric learning: the blue and orange curves respectively show the transformations applied to intraclass and interclass distances. The squares along the horizontal axis present the original distribution of distances, while the stars along the vertical axis shows the distribution of transformed distances. (**a**) Additive. (**b**) Multiplicative. (**c**) Concave vs convex

The scale parameter σ is set based on the average of all distances $\{\Delta_{ij}\}$ to avoid explosions of the exponential terms. The constant Δ_0 is positive and chosen so that the values of $f^{\not\in}$ for the smallest unsupervised distances may be lower than those of $f^{\in}$ for the highest unsupervised distances, while still ensuring that $f^{\not\in}(\Delta_{ij}) > f^{\in}(\Delta_{ij})$ for any value of Δ_{ij} (Fig. 6.3).

6.3.4 Similarities Learning

Some DR techniques, such as the Neighbourhood Embedding family, rely on measures of similarities between points instead of distances. In this case, the approach of metric learning may be adapted either at the level of distances, then processed to obtain similarities, as for S-NeRV, or directly at the level of similarities. In the latter case, the transformation searches to amplify intraclass similarities compared with interclass similarities. This is done for the supervised version of heavy-tailed tSNE [202], using transformed data space membership degrees β'_{ij}. They are computed as

$$\beta'_{ij} \triangleq (1 - \epsilon)\beta_{ij} + \epsilon q_{ij},$$

with ϵ being the level of supervision in $[0, 1]$ and q_{ij} a normalized class-community measure. This measure is computed as:

$$q_{ij} \triangleq \begin{cases} \frac{1_{L_i = L_j}}{|C_{L_i}|}, & \text{if } L_i = L_j \text{ (or } i, j \in \mathcal{S}^{\in}), \\ 0, & \text{if } L_i \neq L_j \text{ (or } i, j \in \mathcal{S}^{\not\in}), \end{cases}$$

where $|C_{L_i}|$ is the cardinal of the class of point i allowing a row-wise normalization (as for the membership degrees). In this method, points with missing labels are treated as if they were ofe a different class than any other point.

The Random-Forest Potential of Heat-diffusion for Affinity-based Transition Embedding (**RF-PHATE**) [151] introduces a supervised similarity measure used as input of the unsupervised **PHATE** algorithm [130]. In order to compute the "random forest proximities", a random forest classifier is trained on the data. Then, the proximity between two points is obtained as the proportion of trees in the random forest for which the two points under consideration are in the same leaf node. Thus, this measure of similarity tends to discard the differences between points for variables that do not impact the classification.

6.3.5   Metric Learning Limitations

The approach of metric learning provides an efficient way to emphasize class separation. Thus, it leads to embeddings with a high class purity allowing good performances for classification algorithms in the low-dimensional space. Yet, at the core, this approach distorts the structure of data, which is assessed through the distance matrix. In that regard, metric learning approaches are not well-suited for exploratory analysis of labeled data.

6.4   Class Adaptive Scale

Many DR techniques, including the neighbourhood embedding family, focus on the preservation of the data structure at a specific scale. Class-aware tSNE (**catSNE**) [48] incorporates class-information in **tSNE** by setting that scale parameter depending on the distribution of classes. The proposed method maximizes the scale parameter σ_i for each point i under the constraint that the class purity remains above a user-set threshold $\theta \in [0.5, 1[$. Thus, if a point is within an area populated with points from its class, the size of the neighbourhood to preserve will be set so as to encompass all this area as well as a few points from other classes. In practice, the purity is assessed using membership degrees β_{ij} defined Eq. (5.13) leading to express this constraint for all i as:

$$\sum_{j \in S_i^{\in}} \beta_{ij} > \theta.$$

For some points, such as outliers in areas mainly of another class, this constraint may be impossible to satisfy. In that case, the scale is chosen so as to maximize the class purity. Hence,though this approach of supervision uses class-information to adapt the considered neighbourhood size and to mostly preserve intraclass proximities, it still focuses on retaining the neighbourhood structure.

6.5 Class-Guiding Principle: Classimap

Another way of exploiting class-information, called here class-guiding, is to steer the unavoidable distortions of data structure, so as to maximize class preservation. This amounts to mostly preventing missed neighbours within a same class (class cohesion) and false neighbours between different classes (class distinction), while allowing more tolerance to other types of distortions. In other words, class-guiding permits some tears between classes and gluings within classes if distortions are necessary to embed the dataset in the lower dimensional space. In this respect, it differs from metric learning approaches, that first distort the structure to emphasize class separation, and then apply unsupervised dimensionality reduction, which attempts to avoid all types of distortions without distinction.

This paradigm of class-guided dimensionality reduction is introduced by the Classimap method [116], which is based on the tunable method Local MultiDimensional Scaling (LMDS) [187]. local multidimensional scaling allows the user to choose whether to penalize more the manifold gluing or the manifold tears, through the introduction of a trade-off parameter $\tau \in [0, 1]$. Classimap proposes to use distinct values of this parameter for the different pairwise relations, distinguishing between $\tau^{\in}$ for intraclass and $\tau^{\notin}$ for interclass relations. Those sets $S_i^{\in}$ and $S_i^{\notin}$ of points indices respectively in the same class as i or in a different class than i, are defined by:

$$S_i^{\in} \triangleq \{j \neq i \mid L_j = L_i\} \quad \text{and} \quad S_i^{\notin} \triangleq \{j \neq i \mid L_j \neq L_i\}.$$

Thus, the LMDS stress given by Eq. (5.11) (in Sect. 5.2.1.2) is broken down as follows:

$$\zeta_i^{\text{Classimap}} \triangleq \sum_{j \in S_i^{\in}} (\Delta_{ij} - D_{ij})^2 w_{ij}(\tau^{\in}) + \sum_{j \in S_i^{\notin}} (\Delta_{ij} - D_{ij})^2 w_{ij}(\tau^{\notin}), \quad (6.3)$$

with the weights

$$w_{ij}(\tau) \triangleq \tau w\left(\frac{\Delta_{ij}}{\sigma_i}\right) + (1 - \tau) w\left(\frac{D_{ij}}{s_i}\right).$$

The trade-off values are chosen in order to penalize more tears within a same class and gluings between different classes, which translates by $\tau_{\in} \geqslant \tau_{\notin}$. When $\tau_{\in} = \tau_{\notin}$, Classimap is equivalent to the unsupervised LMDS.

Classimap may also be formulated so as to allow a soft estimation of class-community with a parameter τ_{ij} varying progressively between 0 and 1:

$$\zeta_i^{\text{Classimap}} \triangleq \sum_{j \neq i} (\Delta_{ij} - D_{ij})^2 w_{ij}(\tau_{ij}). \tag{6.4}$$

The class-guided technique tries to steer the convergence of the algorithm towards a solution that could have been reached by penalizing indistinctly the two types of distortions, but ensuring that the necessary distortions of the structure for the obtained solution are the most compatible with class preservation (cohesion and separation). We may note that when a mapping perfectly preserving the data structure exists, both the unsupervised and class-guided technique are supposed to converge to the same solution.

6.6   Class-Guided Neighbourhood Embedding

Class-Guided Neighbourhood Embedding (CGNE) adapts the intuitions of class-guiding introduced by Classimap to supervise Neighbourhood Embedding techniques. Compared with LMDS, which is a metric MDS method, Neighbourhood Embedding techniques benefit from an interesting theoretical framework for map interpretation, which is the neighbourhood retrieval perspective [189]. They also offer practical qualities such as the shift-invariance property which provides robustness to the curse of dimensionality [107]. Within the family of Neighbourhood Embedding methods, the best-suited for adapting the class-guiding approach are NeRV and JSE which, similarly to LMDS, incorporate a trade-off parameter allowing to balance false and missed neighbours. This leads to the definition of ClassNeRV [36] and ClassJSE.

Figure 6.4 illustrates the interest of the class-guiding approach for a three-dimensional toy dataset. This dataset is constituted of three Gaussian clusters and contains four classes. On the one hand, the unsupervised mapping (Fig. 6.4b) correctly represents the three clusters, but induces overlap between the orange and blue classes, which were linearly separable in the data space. Indeed, the technique not accounting for class-information induces random distortions of the structure, which happen to affect the representation of classes. On the other hand, the supervised mapping (Fig. 6.4c) over-separates the classes, thus distorting the data structure. Indeed, it represents five different clusters, disregarding the connection between the orange and blue classes and the mixing of the pink and green classes. Relying on the class-guiding approach, ClassNeRV preserves the three clusters structure (class cohesion) while avoiding overlaps between the blue and orange classes (class distinction) thanks to the additional class-information. As such, ClassNeRV provides a good support for visual exploration of labelled data.

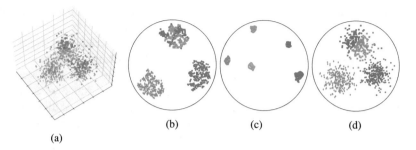

Fig. 6.4 Embeddings of a toy dataset constituted of three Gaussian clusters, the first purely of the blue class, the second linearly separated between the blue and orange class and the third with mixed pink and green classes. (**a**) 3D clusters. (**b**) tSNE. (**c**) S-UMAP. (**d**) ClassNeRV

6.6.1 *ClassNeRV Stress*

In order to introduce supervision into **NeRV**, we propose to use a different trade-off for intra-class and inter-class relations in **NeRV** stress. As a reminder this stress, introduced Sect. 5.3.3, is given by:

$$\zeta_i^{\text{NeRV}} = \tau \mathcal{D}_{\text{KL}}(\beta_i, b_i) + (1 - \tau)\mathcal{D}_{\text{KL}}(b_i, \beta_i),$$

where $\beta_i \triangleq (\beta_{ij})_{j \neq i}$ and $b_i \triangleq (b_{ij})_{j \neq i}$ are the neighbourhood membership degree defined as:

$$\beta_{ij} = \frac{\exp\left(-\Delta_{ij}^2/2\sigma_i^2\right)}{\sum_{k \neq i} \exp\left(-\Delta_{ik}^2/2\sigma_i^2\right)} \quad \text{and} \quad b_{ij} = \frac{\exp\left(-D_{ij}^2/2s_i^2\right)}{\sum_{k \neq i} \exp\left(-D_{ik}^2/2s_i^2\right)}. \tag{6.5}$$

In the following, we consider scale parameters σ_i set by perplexity and $s_i = \sigma_i$ as detailed in Sect. 5.3.2. A naive approach to this problem consists in breaking down the classical Kullback–Leibler divergence in the following way:

$$\mathcal{D}_{\text{KL}}(\beta_i, b_i) = \sum_{j \in \mathcal{S}_i^{\in}} \beta_{ij} \log\left(\frac{\beta_{ij}}{b_{ij}}\right) + \sum_{j \in \mathcal{S}_i^{\notin}} \beta_{ij} \log\left(\frac{\beta_{ij}}{b_{ij}}\right). \tag{6.6}$$

However, with this formulation, the sub-stresses are not necessarily positive. Indeed, the classical Kullback–Leibler divergence is guaranteed to be positive only when its arguments are probability distributions, namely mass functions summing to one such as β_i and b_i. Yet, $\beta_i^{\in} \triangleq (\beta_{ij})_{j \in \mathcal{S}_i^{\in}}$ and the equivalently defined $\beta_i^{\notin}$, $b_i^{\in}$ and $b_i^{\notin}$ are *partial distributions*, defined here as mass functions summing to strictly less than one. We may note that the partial distributions $\beta_i^{\in}$, $\beta_i^{\notin}$, $b_i^{\in}$ and $b_i^{\notin}$ may sum to one only in those two pathological configurations: all points are of the same class, or they are all of different classes. In both cases, the labels have no supervision effect.

As a result, the individual sub-stresses of Eq. (6.6) may reach negative values. Yet, for the ideal case of perfectly preserved neighbourhoods, namely $\beta_{ij} = b_{ij}$ for all pairs (i, j), these sub-stresses are equal to zero. Thus, the minimum value is reached for a suboptimal configuration, which is not the desired behaviour for a stress function. To avoid this short-coming, the break-down of the classical Kullback–Leibler divergence may be reformulated with generalized Kullback–Leibler sub-terms:

$$\mathscr{D}_{\mathrm{KL}}\,(\beta_i, b_i) = \sum_{j \in S_i^{\in}} \beta_{ij} \log\left(\frac{\beta_{ij}}{b_{ij}}\right) + b_{ij} - \beta_{ij} + \sum_{j \in S_i^{\notin}} \beta_{ij} \log\left(\frac{\beta_{ij}}{b_{ij}}\right) + b_{ij} - \beta_{ij},$$

(6.7)

using the fact that $\sum_{j \neq i} \beta_{ij} = \sum_{j \neq i} b_{ij} = 1$.

The generalized Kullback–Leibler divergence for the partial intraclass distributions is denoted as $\mathscr{D}_{\mathrm{gKL}}(\beta_i^{\in}, b_i^{\in})$ and given by:

$$\mathscr{D}_{\mathrm{gKL}}(\beta_i^{\in}, b_i^{\in}) \triangleq \sum_{j \in S_i^{\in}} \beta_{ij} \log\left(\frac{\beta_{ij}}{b_{ij}}\right) + b_{ij} - \beta_{ij},$$

with a symmetric definition for the interclass divergence $\mathscr{D}_{\mathrm{gKL}}(\beta_i^{\notin}, b_i^{\notin})$.

The generalized Kullback–Leibler divergence is the Bregman divergence [24] whose generator function is the negative Shannon entropy (see Annex B.1):

$$-H(\beta_i^{\in}) \triangleq \sum_{j \in S_i^{\in}} \beta_{ij} \log \beta_{ij}.$$

This ensures the positivity even for partial distributions $\beta_i^{\in}$, $b_i^{\in}$, $\beta_i^{\notin}$ and $b_i^{\notin}$. We may note that for non-partial distributions, the generalized and classical Kullback–Leibler divergences are equivalent, meaning that $\mathscr{D}_{\mathrm{gKL}}(\beta_i, b_i) = \mathscr{D}_{\mathrm{KL}}(\beta_i, b_i)$.

Linear combination of the break-down of Eq. (6.7) for soft-recall $\mathscr{D}_{\mathrm{KL}}(\beta_i, b_i)$ and soft-precision $\mathscr{D}_{\mathrm{KL}}(b_i, \beta_i)$ using different trade-off parameters for the intraclass and interclass terms, then leads to **ClassNeRV** stress in its expanded form:

$$\zeta_i^{\mathrm{ClassNeRV}} \triangleq \tau^{\in} \mathscr{D}_{\mathrm{gKL}}(\beta_i^{\in}, b_i^{\in}) + (1 - \tau^{\in}) \mathscr{D}_{\mathrm{gKL}}(b_i^{\in}, \beta_i^{\in})$$
$$+ \tau^{\notin} \mathscr{D}_{\mathrm{gKL}}(\beta_i^{\notin}, b_i^{\notin}) + (1 - \tau^{\notin}) \mathscr{D}_{\mathrm{gKL}}(b_i^{\notin}, \beta_i^{\notin})$$

(6.8)

with the intraclass trade-off $\tau^{\in} \in [0, 1]$ and the interclass trade-off $\tau^{\notin} \in [0, 1]$ so that $\tau^{\in} \geqslant \tau^{\notin}$.

The four terms of the expanded stress of Eq. (6.8) may be grouped horizontally, which gives a more compact expression of **ClassNeRV** stress in terms of mixtures on partial distributions:

$$\zeta_i^{\text{ClassNeRV}} = \mathcal{M}_{\text{gKL},1}(b_i^{\in}, \beta_i^{\in}, \tau^{\in}) + \mathcal{M}_{\text{gKL},1}(b_i^{\notin}, \beta_i^{\notin}, \tau^{\notin}), \qquad (6.9)$$

with the generalized type 1 mixture between intraclass partial distributions:

$$\mathcal{M}_{\text{gKL},1}(b_i^{\in}, \beta_i^{\in}, \tau^{\in}) \triangleq \tau^{\in} \mathscr{D}_{\text{gKL}}(\beta_i^{\in}, b_i^{\in}) + (1 - \tau^{\in})\mathscr{D}_{\text{gKL}}(b_i^{\in}, \beta_i^{\in}).$$

The equivalent is defined for interclass partial distributions. In this reformulation, the first term penalizes errors in the preservation of intraclass neighbourhoods, whereas the second penalizes errors in the preservation of interclass-neighbourhoods. The use of $\tau^{\in} > \tau^{\notin}$ allows to introduce supervision by penalizing more missed neighbours within classes and false neighbours between classes. Conversely, $\tau^{\in} = \tau^{\notin} = \tau$ leads back to the unsupervised **NeRV** stress with trade-off τ, as may be seen by grouping vertically the terms in Eq. (6.8).

We may note that intraclass soft recall $\mathscr{D}_{\text{gKL}}(\beta_i^{\in}, b_i^{\in})$, intraclass soft precision $\mathscr{D}_{\text{gKL}}(b_i^{\in}, \beta_i^{\in})$ and their interclass counterparts may be linked with class-aware map quality indicators intraclass and interclass recall and precision (see Annex B.2). In the following, we also refer to soft recall and soft precision restricted to intraclass or interclass relations as $\tilde{\mathscr{R}}^{\in}, \tilde{\mathscr{P}}^{\in}, \tilde{\mathscr{R}}^{\notin}$ or $\tilde{\mathscr{P}}^{\notin}$.

As for optimization, the stress of **CGNE** methods (**ClassNeRV** and **ClassJSE**) is optimized with the **BFGS** quasi-Newton algorithm [138]. It also uses a **cMDS** initialization [177] and a multi-scale optimization paradigm [110]. Those different elements are detailed in Sect. 7.1.

6.6.2 Flexibility of the Supervision

We consider here the impact of parameters $\tau^{\in}$ and $\tau^{\notin}$ on the result of **ClassNeRV**, as shown Figs. 6.5 and 6.6. For more intuitive interpretation, the method may be re-parametrized using the average trade-off τ^* and the level of supervision ϵ expressed by:

$$\tau^* = \frac{\tau^{\in} + \tau^{\notin}}{2} \text{ and } \epsilon = \frac{\tau^{\in} - \tau^{\notin}}{2}.$$

With that reformulation, the trade-off parameters $\tau^{\in}$ and $\tau^{\notin}$ may be simply obtained as:

$$\tau^{\in} = \tau^* + \epsilon \text{ and } \tau^{\notin} = \tau^* - \epsilon$$

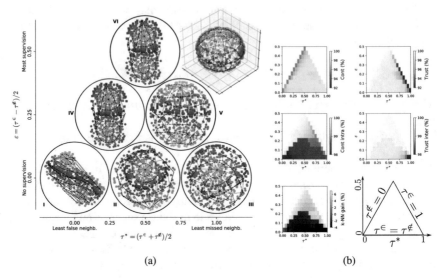

(a)   (b)

Fig. 6.5 Sensitivity of ClassNeRV to the trade-off parameters with the sphere dataset. Results are presented in the space parametrized by τ^* and ϵ. Panel (**a**) shows six ClassNeRV maps with a posteriori positioned lines representing the parallels and meridians of the sphere. Heatmaps of panel (**b**) show the evolution of quality indicators with those parameters for 121 different maps. The colour bars of those indicators are restricted to the smallest integer interval containing the 5th and 95th percentile of the associated indicators value. A common range is used for continuity $\check{\mathscr{C}}(\kappa)$ and trustworthiness $\check{\mathscr{T}}(\kappa)$ (abbreviated Cont and Trust) and their class-aware equivalent $\check{\mathscr{C}}^{\in}$ and $\check{\mathscr{C}}^{\notin}$ (Cont intra, Trust inter). All indicators are computed for $\kappa = 32$, which is also the value of the perplexity κ^* used for ClassNeRV. The triangular shape of the domain is explained by the constraints that $\tau^{\in}$ and $\tau^{\notin}$ must be between 0 and 1, and $\tau^{\in} \geqslant \tau^{\notin}$. (**a**) Maps. (**b**) Indicators

The goal of that re-parametrization is to partially decouple the global balance between false and missed neighbours and the level of supervision. We consider the effect of those parameters for the sphere dataset with two classes linearly separated by the equator plane, and for a random sub-sampling of the high-dimensional digits dataset (see Sect. 1.1.7), each containing 512 data instances. Note that those datasets are both non-mappable in a two-dimensional embedding space.

Figure 6.5a presents several maps of the sphere dataset obtained with different values of ClassNeRV hyper-parameters. The maps I, II and III ($\epsilon = 0$) correspond to the unsupervised case and are equivalent to NeRV with a trade-off τ^*. When the parameter τ^* increases from 0 to 1, missed neighbours are more and more penalized, while false neighbours are less and less penalized. Thus, for $\tau^* = 0$, the sphere is approximately torn along a meridian and unfolded, leading to intraclass missed neighbours. Inversely, for $\tau^* = 1$, which corresponds to SNE, the sphere is squashed, inducing interclass false neighbours. In the intermediate case of $\tau^* = 0.5$ the blue hemisphere is well preserved, while the orange hemisphere is spread around it.

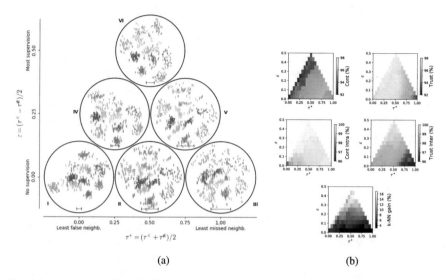

Fig. 6.6 Sensitivity of ClassNeRV to the trade-off parameters with the **digits** dataset. All details, except the datasets are common to Fig. 6.5. A scale is also added on the map, representing the data radius $\Delta_{\max}/2$. (**a**) Maps. (**b**) Indicators

Adding an intermediate supervision ($\epsilon = 0.25$) leads to maps IV and V. Similarly to the unsupervised cases, the sphere is torn or squashed, but the class-information provided to ClassNeRV allows to choose meaningful distortions with regard to the classes. For $\tau^* = 0.25$, missed neighbours are favoured on average, inducing a tear along the equator. Conversely, for $\tau^* = 0.75$, false neighbours are globally less penalized, so that the collapse of the sphere preserves the separation of the two hemispheres. In the maximally supervised case shown by map VI ($\epsilon = 0.5$), corresponding to the top map, only intraclass false neighbours and interclass missed neighbours are penalized. This induces a map very similar to the one obtained with $\tau^* = 0.25$ and $\epsilon = 0.25$.

The heatmaps of Fig. 6.5b show the evolution of quality indicators for many different maps of the **sphere**. The trustworthiness $\check{\mathscr{T}}$ and continuity $\check{\mathscr{C}}$ indicators demonstrate that, for a given ϵ, the amount of false neighbours increases with τ^*, while the amount of missed neighbours decreases. Furthermore, the class-aware indicators $\check{\mathscr{T}}^{\notin}$ and $\check{\mathscr{C}}^{\in}$ show that the amount of distortions specifically affecting classes decreases when the level of supervision ϵ increases. This decrease also leads to a higher class purity, as shown by the κ-NN gain indicator. Thereby, this quantitative analysis confirms the trends observed on the maps.

Roughly similar results are presented Fig. 6.6 for the **digits** dataset. In the unsupervised case ($\epsilon = 0$), the scales added on the maps indicate that when τ^* decreases (penalizing less missed neighbours than false neighbours), the map is more stretched globally. When the level of supervision ϵ increases, the classes appear more separated. As for the heatmaps of quality indicators, they show similar trends for the digits than for the sphere dataset. We may however note that, as opposed to the sphere, the digits dataset has a high intrinsic dimensionality. As such, it requires more distortion of the structure in order to be mapped in a two-dimensional space. Thus, for the sphere, the terms mostly impact the initial breaking of the global topological structure, whereas for the digits, the balancing of the terms defines the complex equilibrium state reached at the end of the optimization. This may contribute to explain the higher difference between maps of the digits data set, compared with maps of the sphere.

6.6.3   Ablation Study

In order to examine more thoroughly the individual impact of each of the four terms of ClassNeRV stress, an ablation study is conducted. Ablation studies consist in removing one element of an algorithm and observing the variation of the result that is obtained in this way. To perform our study, we consider the following reformulation of ClassNeRV stress, which allows more degrees of freedom in the balancing of the four terms. That reformulation is:

$$\zeta_i^{\text{ClassNeRV}} = w_{\tilde{\mathcal{R}}}^{\in} \mathscr{D}_{\text{gKL}}(\beta_i^{\in}, b_i^{\in}) + w_{\tilde{\mathcal{P}}}^{\in} \mathscr{D}_{\text{gKL}}(b_i^{\in}, \beta_i^{\in})$$
$$+ w_{\tilde{\mathcal{R}}}^{\notin} \mathscr{D}_{\text{gKL}}(\beta_i^{\notin}, b_i^{\notin}) + w_{\tilde{\mathcal{P}}}^{\notin} \mathscr{D}_{\text{gKL}}(b_i^{\notin}, \beta_i^{\notin}),$$

(6.10)

where weights $w_{\tilde{\mathcal{R}}}^{\in}$, $w_{\tilde{\mathcal{P}}}^{\in}$, $w_{\tilde{\mathcal{R}}}^{\notin}$ and $w_{\tilde{\mathcal{P}}}^{\notin}$ all range from 0 to 1. In the ablation study, those weights are binary. Thus, the set of weights leading to a given map may be summarized visually by a four panels view, as illustrated in Fig. 6.7. In this Figure, the set of weights corresponds to the case of unsupervised SNE (i.e. ClassNeRV for $\tau^* = 1$ and $\epsilon = 0$).

First, we consider the impact of removing only one of the four terms in Eq. (6.10). This means performing an ablation of terms of the balanced NeRV stress (i.e. ClassNeRV with $\tau^{\in} = \tau^{\notin} = 0.5$). Figure 6.8 presents the results of this ablation

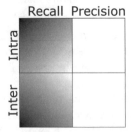
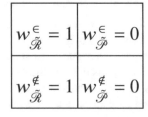

Fig. 6.7 Four panels representation of the weights in ClassNeRV ablation study. Black and white panels respectively represent weights equal to 1 or 0

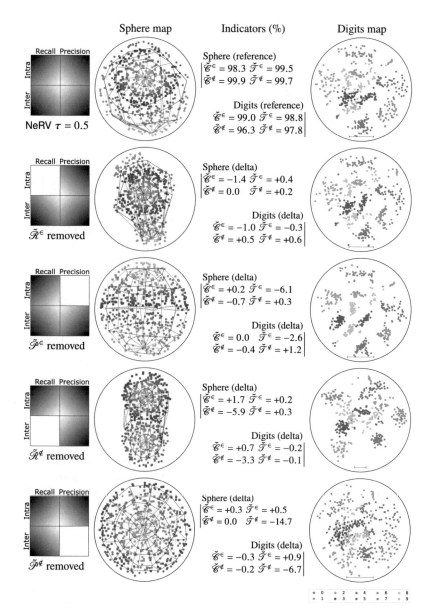

Fig. 6.8 Ablation study for the unsupervised case of ClassNeRV ($\tau^\in = \tau^\notin = 0.5$). The four panel views on the left side encode the weights set to zero or one. Indicators are computed for $\kappa = 32$, and given either by their absolute value or by their difference with respect to the reference case (delta)

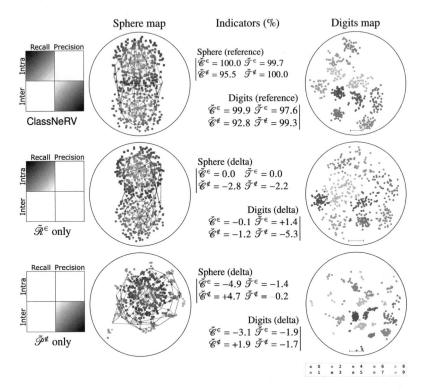

Fig. 6.9 Ablation study for the most supervised case of ClassNeRV ($\tau^{\in} = 1$, $\tau^{\notin} = 0$). See the caption of Fig. 6.8 for more details

for the **sphere** and **digits** datasets. Then, we consider the maps obtained with only one term of the stress. This is equivalent to performing an ablation study on the most supervised case of **ClassNeRV** ($\tau^{\in} = 1$ and $\tau^{\notin} = 0$), or on its anti-supervised version ($\tau^{\in} = 0$ and $\tau^{\notin} = 1$). The results of those studies are presented Figs. 6.9 and 6.10. The class-aware quality indicators are added to quantify the amount of intra-class and inter-class missed and false neighbours for each map. In a given ablation study, those indicators are expressed as deltas, that is, as the deviation between a map with ablation of a term and its reference map.

With this presentation, the indicators show a strong correlation between soft recall and continuity, as well as between soft precision and trustworthiness, both for the restrictions to intraclass and to interclass relations. Indeed, the removal of a term of the stress systematically engenders a significant drop in the value of the associated quality indicator, with a difference of at least 1% in absolute value. In the same time the other indicators are also impacted, either leading to an increase or decrease of their values. An increase could be explained by the relaxation of the constraint imposed by the removed term, which would in turn allow to improve other aspects of the map quality. This seems to be the case when removing intraclass soft recall $\tilde{\mathscr{R}}^{\in}$ from the supervised **ClassNeRV** ($\tilde{\mathscr{P}}^{\in}$ only in Fig. 6.9). In that case, the

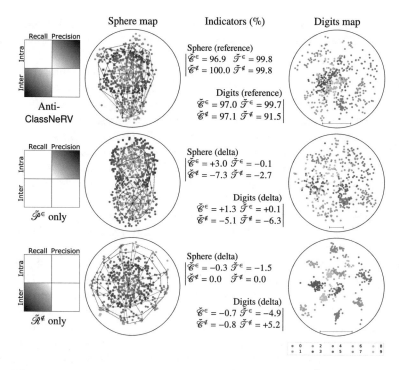

Fig. 6.10 Ablation study for the anti-supervised ClassNeRV ($\tau^\in = 0$, $\tau^\notin = 1$). See the caption of Fig. 6.8 for more details

relaxation of the constraint imposed on intraclass continuity $\check{\mathscr{C}}^\in$ leads to an increase of interclass continuity $\check{\mathscr{C}}^\notin$. Conversely, a slight decrease of a quality indicator could mean that the missing term helped to obtain a meaningful structure of the data, thus also contributing to improve the aspect of map quality assessed by that indicator. In the same example as above, the intraclass soft recall provided an important part of the global structure by ensuring class cohesion. Thus, removing it does not decrease only its associated indicator of interclass continuity $\check{\mathscr{C}}^\notin$, but also the trustworthiness terms $\check{\mathscr{T}}^\in$ and $\check{\mathscr{T}}^\notin$. These quantitative results are illustrated by the maps.

Maps of Fig. 6.8 show how the balance between the four terms is affected by removing one term. Focusing on the maps of the sphere, we see that removing the intraclass soft recall $\tilde{\mathscr{R}}^\in$ leads to tearing the orange class, while preserving the equatorial frontier. Ablation of the intraclass soft precision $\tilde{\mathscr{P}}^\in$ induces a collapse of hemispheres onto themselves. Conversely, the removal of interclass soft recall $\tilde{\mathscr{R}}^\notin$ allows the sphere to be torn along the equator. As for interclass soft precision $\tilde{\mathscr{P}}^\notin$, its ablation causes the hemispheres to be squashed onto each other. For maps of the digits dataset, ablation of intraclass soft recall $\tilde{\mathscr{R}}^\in$ leads to more points lying out of their class than in NeRV, whereas ablation of soft precision $\tilde{\mathscr{P}}^\in$ induces more clumped classes. Removing interclass soft-recall $\tilde{\mathscr{R}}^\notin$ allows more separation of classes, which looks rather similar to the previous case, but with more spatial spreading of the points. Finally, without soft precision $\tilde{\mathscr{P}}^{\notin}$, strong overlap occurs between classes.

For Figs. 6.9 and 6.10, the maps show what is preserved by each term considered individually. For the terms of supervised ClassNeRV (Fig. 6.9), the respective role of the intraclass soft recall and interclass soft precision appear clearly. On the one hand, when considered alone intraclass soft recall $\tilde{\mathscr{R}}^{\in}$ preserves the cohesion of each class. For the spheres this leads to two well-represented hemispheres with a minimal junction between the two, while for the digits, classes are represented as coherent blocks slightly overlapping with their neighbouring classes. On the other hand, interclass soft precision $\tilde{\mathscr{P}}^{\notin}$ prevents overlaps between different classes, though allowing significant tears within a same class. We may also note an effect of that soft precision term on interclass cohesion for the sphere. Indeed, this term used alone is sufficient to maintain the equatorial frontier. Its removal also induces more tearing along that line.

Figure 6.10 presents the ablation for the anti-supervised case of ClassNeRV. This case combines the intraclass soft precision and interclass soft recall. Intraclass soft precision $\tilde{\mathscr{P}}^{\in}$ alone avoids the collapse of classes. Thus, it leads to two clear hemispheres, with a loose frontier, and to spread out digits classes strongly overlapping with each other. Interclass soft recall $\tilde{\mathscr{R}}^{\notin}$ ensures that interfaces between classes are not torn. Hence, the map of the sphere preserves the equatorial frontier, with the orange class regularly spread around the blue. For the digits, the classes are relatively densely grouped, with some outlying points placed close to another class. The anti-supervised case mixes those two, with the hemispheres being well-connected and regions of the orange hemisphere also appearing clearly, such as the polar region represented at the bottom of the map. On the other hand, the digits classes appear significantly entangled, with some torn classes.

6.6.3.1   Comparison with Other Dimensionality Reduction Methods

In this Section, we compare the results of ClassNeRV to other unsupervised and supervised Dimensionality Reduction techniques. This comparison focuses on the previously considered sphere and digits datasets, but it also extends to the case of the Isolet 5 dataset detailed in Sect. 6.6.4. The results of ClassNeRV are given for $\tau^* = 0.5$ and three levels of supervision: no supervision $\epsilon = 0$ (i.e. NeRV), medium supervision $\epsilon = 0.3$ and maximum supervision $\epsilon = 0.5$. The maps of ClassNeRV are presented in Figs. 6.11 and 6.14, while those obtained with other methods are shown in Figs. 6.12 and 6.15. Quantitative comparisons are presented in Figs. 6.13 and 6.16. In these indicators, the average $(\check{\mathscr{T}} + \check{\mathscr{C}})/2$ of trustworthiness and continuity indicates the overall preservation of the data structure. Its class-aware counterpart $(\check{\mathscr{T}}^{\notin} + \check{\mathscr{C}}^{\in})/2$ assesses the specific preservation of classes structure. Finally, the level of class purity is measured with the κ-NN gain. Note that we consider here the most supervised version of ClassNeRV ($\tau^* = 0.5, \epsilon = 0.5$). Yet,

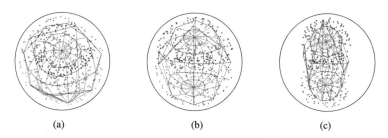

Fig. 6.11 ClassNeRV maps of the sphere with $\tau^* = 0.5$ and different levels of supervision: $\epsilon = 0$ (no supervision), $\epsilon = 0.3$ (medium supervision) and $\epsilon = 0.5$ (maximum supervision). (**a**) No supervision (NeRV). (**b**) Medium supervision. (**c**) Maximum supervision

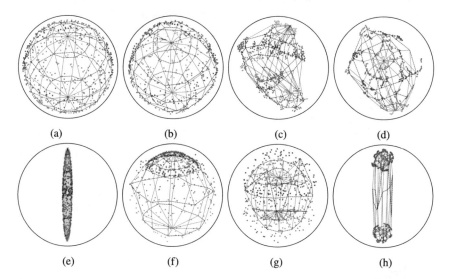

Fig. 6.12 Maps of the sphere obtained by other DR techniques (both unsupervised and supervised). (**a**) PCA (**b**) Isomap. (**c**) tSNE. (**d**) UMAP. (**e**) NCA. (**f**) S Isomap. (**g**) Classimap. (**h**) S-UMAP

a different trade-off between preservation of classes and structure could be reached by decreasing the level of supervision.

For the **sphere** dataset, the maps obtained by the unsupervised methods (Figs. 6.12a–d) either fail to preserve classes cohesion or distinction. Indeed, PCA and Isomap, which rely on a linear projection (either from the three-dimensional space, or from the kernel space of geodesic distances), glue the classes together, losing class distinction. Conversely, the non-linear methods tSNE and UMAP tear them in several pieces, losing class cohesion. As for supervised methods (Figs. 6.12e–h), they manage to maintain class cohesion and distinction with the two hemispheres represented in one well-separated connected component. Yet, most of them distort the data structure. NCA and S-Isomap, which also rely on a linear projection step in their algorithm, both collapse the classes onto themselves.

Furthermore, with NCA the sphere is compressed along its parallels, while with S-Isomap, the blue hemisphere is compressed compared with the orange class. Conversely, Classimap and S-UMAP perform non-linear transformations of the hemispheres, leading to better local representations. However, S-UMAP strongly over-separates the two hemispheres, distorting the equatorial interface. Conversely, the supervised ClassNeRV in Figs. 6.11b and c both maintain class cohesion and distinction. Note that, in the maximally supervised case, both hemispheres are represented almost independently with a preservation of a small boundary region, as opposed to the medium supervision case, for which the equatorial frontier is less torn.

The quality indicators (Fig. 6.13) show that ClassNeRV is comparable with most unsupervised methods in terms of structure preservation $(\check{\mathcal{T}} + \check{\mathcal{C}})/2$ with slightly worse results than NeRV (especially for the maximum supervision). Yet, it leads to better performances for the preservation of the classes structure, measured

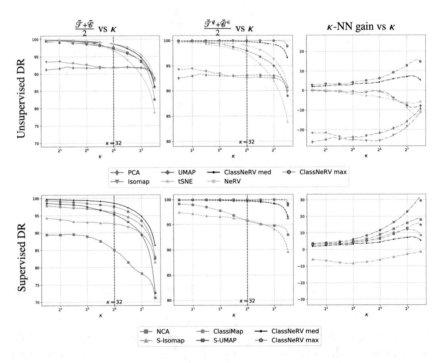

Fig. 6.13 Quantitative comparison of ClassNeRV and other DR techniques for the sphere dataset. These techniques are divided into unsupervised an supervised techniques. Indicators are presented as functions of the neighbourhood size κ, which varies from 1 to $N/2$. The scale $\kappa = 32$ used when presenting scalar indicators and corresponding to the perplexity $\kappa^\star$ is shown by the black dashed line. Indicator $(\check{\mathcal{T}} + \check{\mathcal{C}})/2$ accounts for the level of structure preservation, while $(\check{\mathcal{T}}^{\notin} + \check{\mathcal{C}}^{\in})/2$ shows the specific preservation of the structure of classes. The κ-NN gain indicates the difference of class purity between the map and the data space (a positive value meaning an increase of purity in the map)

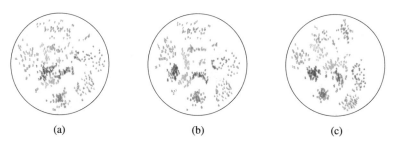

(a) (b) (c)

Fig. 6.14 ClassNeRV maps of the digits with $\tau^* = 0.5$ and different levels of supervision: $\epsilon = 0$ (no supervision), $\epsilon = 0.3$ (medium supervision) and $\epsilon = 0.5$ (maximum supervision). (**a**) No supervision (NeRV). (**b**) Medium supervision. (**c**) Maximum supervision

by $(\check{\mathscr{T}}^{\notin} + \check{\mathscr{C}}^{\in})/2$. It also induces a higher class purity in the map as assessed by the κ-NN gain. These results are coherent with the maps of the **sphere** and with the fact that unsupervised methods focus on the preservation of the data structure, without accounting for class-information.

However, compared with other supervised methods, ClassNeRV leads to the best preservation of structure. We may note that S-UMAP performs relatively well at small scales κ but suffers from more distortion at larger scales, probably due to the over-separation of the hemispheres. In terms of class preservation, the maximally supervised ClassNeRV is similar to S-UMAP, which is the other best supervised method for that characteristic. Finally, considering the κ-NN gain, S-UMAP and NCA show higher class purity than the maximally supervised ClassNeRV. This is coherent with the different purpose of these methods which is to maximize class separation.

For the **digits** dataset (mapped by ClassNeRV in Fig. 6.14), the maps obtained with PCA (Fig. 6.15), NCA (Fig. 6.15) and Isomap (Fig. 6.15) present many class overlaps, probably due to their linear projection step. Inversely, tSNE (Fig. 6.15) and UMAP (Fig. 6.15) manage to clearly separate the classes, despite being unsupervised. These methods also show interfaces between some classes. For the supervised S-Isomap (Fig. 6.15) and S-UMAP (Fig. 6.15), which rely on metric learning, the classes seem to be over-separated. Finally, Classimap (Fig. 6.15) represents each class as a rather spread out cluster. For this dataset, NeRV map (Fig. 6.14a) contains several classes overlap and splits. ClassNeRV with medium supervision (Fig. 6.14b) increases the class distinction, while also preserving some similarities between digits of different classes. Finally, the maximally supervised ClassNeRV (Fig. 6.14c) increases the class cohesion, with the 1s and 9s regrouped in one cluster, showing a little fewer interclass proximities.

Indicators presented Fig. 6.16 show that ClassNeRV performances are similar to those of the unsupervised NeRV, UMAP and tSNE for structure preservation, and slightly better for class preservation. We may however notice worse results than NeRV at higher scales κ (especially with the maximum supervision). Compared with supervised methods, ClassNeRV induces a better structure preservation than

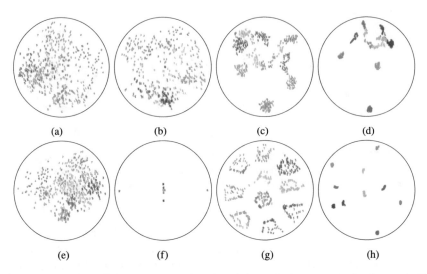

Fig. 6.15 Maps of the digits obtained by other DR techniques (both unsupervised and supervised). (**a**) PCA. (**b**) Isomap. (**c**) tSNE. (**d**) UMAP. (**e**) NCA. (**f**) S-Isomap. (**g**) Classimap. (**h**) S-UMAP

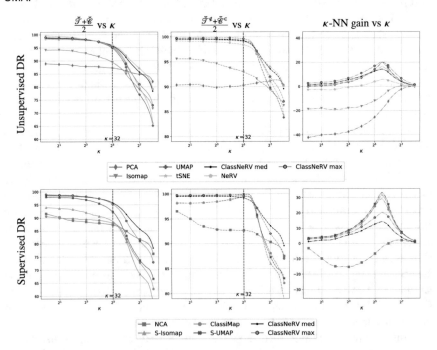

Fig. 6.16 Quantitative comparison of ClassNeRV and other DR techniques for the digits dataset (with true labels). Technicalities are similar to the caption of Fig. 6.13

other methods and a class preservation equivalent to that of **S-UMAP**, whose performances decrease at high scale κ. In terms of class purity, the map of **ClassNeRV** leads to more purity than the unsupervised methods, but less than most of the other supervised methods.

As a conclusion, the maximally supervised **ClassNeRV** focuses on class cohesion and distinction, avoiding irrelevant classes overlaps and splits. In doing so, it tends to provide a map with clearly separated classes, disregarding some interclass relations and slightly affecting the preservation of the data structure. However, this impact on the structure is far lower than that of other supervised methods. For the considered indicators, **ClassNeRV** with medium supervision appears to be systematically between **NeRV** and the maximally supervised **ClassNeRV**. Yet, it is most often closer from the method with the highest indicator value. Hence, that intermediate solution may provide a good trade-off when attempting to preserve both the classes and the structure.

6.6.4 *Isolet 5 Case Study*

We then consider the case of the **Isolet 5** dataset introduced Sect. 1.1.7. For that dataset, we may have some intuitions about the oral similarities between different letters (i.e. classes), especially relying on phonetic notations. Yet, it is more difficult to get an idea of the similarity between two specific instances of different classes, than it is for the **sphere** and **digits** dataset. Indeed, for the **sphere**, the three-dimensional structure represented by the lines in the map is known, while for the **digits**, images may be compared to assess intuitively the similarities. Hence, the **Isolet 5** dataset is characterized by its purity matrix (Fig. 6.17). This matrix might be interpreted as the confusion matrix of a κ-NN classifier in the data space returning soft labels. Each element (i, j) of that matrix indicates the average proportion of points of class j in the κ-neighbourhood of points of class i (for $\kappa - 32$). Thus, it gives an indication of the interfaces or overlaps between classes in the data space. The classes are re-ordered so as to maximize the block structure of that matrix. Plain lines are added to highlight the blocks whose elements are higher than 20% and dotted lines for blocks with elements higher than 10%.

For that dataset, the map obtained with the maximally supervised **ClassNeRV** is presented Fig. 6.18. In this map, many classes appear well-separated from the others, such as the Q [kjuː], U [juː], H [eɪtʃ] or Y [waɪ]. Alternately, some classes are represented with strong common interfaces. This is the case of classes C [siː] and Z [ziː], J [dʒeɪ] and K [keɪ], M [ɛm] and N [ɛn], L [ɛl] and O [oʊ] or F [ɛf] and S [ɛs]. Finally, classes B [biː], D [diː], E [iː], G [dʒiː], P [piː], T [tiː] and V [viː] strongly overlap in a common region of the map. These groups are coherent with blocks identified in the purity matrix. Hence, **ClassNeRV** map support the discovery of those relations between the classes. Considering the higher scale structure, we may identify similarities between better-separated groups, such as A [eɪ] and H [eɪtʃ], Q [kjuː] and U [juː], S [ɛs], F [ɛf] and X [ɛks], R [ɑr], I [aɪ] and Y [waɪ]. We may

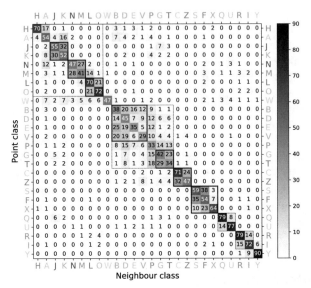

Phonetic: H [eɪtʃ], A [eɪ], J [dʒeɪ], K [keɪ], N [ɛn], M [ɛm], L [ɛl], O [oʊ], W [ˈdʌbəljuː], B [biː], D [diː], E [iː], V [viː], P [piː], G [dʒiː], T [tiː], C [siː], Z [ziː], F [ɛf], S [ɛs], X [ɛks], Q [kjuː], U [juː], R [ɑr], I [aɪ], Y [waɪ].

Fig. 6.17 Purity matrix for the Isolet 5 dataset. Elements (i, j) of that matrix indicate the average proportion of neighbours of class j in the κ-neighbourhood of points of class i ($\kappa = 32$). Phonetics of letters are added to help understand classes proximities

note that the similarities between letters seem to be mainly explained by their vowel sound (e.g., [iː], [eɪ], [ɛ], [aɪ]). A secondary effect of the consonant may explain other distinctions. The latter justifies for instance the difference between the group contstituted of the C [siː] and Z [ziː] class and the group B [biː], D [diː], E [iː], G [dʒiː], P [piː], T [tiː] and V [viː]. Moreover, this multi-level structure of the classes revealed by **ClassNeRV** may be seen as a hierarchy.

With a lower supervision (Fig. 6.18), **ClassNeRV** approximately maintain that global class structure, although classes appear less clearly separated. We may for instance notice class splits and overlaps in the group of the A [eɪ], J [dʒeɪ] and K [keɪ] classes. This tendency is even stronger in the unsupervised case (Fig. 6.18). However, those maps map reveal an interesting characteristic of the class W [ˈdʌbəljuː]. This class is spread out with some elements rather close from the group of classes L [ɛl] and O [oʊ] and other positioned near the group of classes Q [kjuː] and U [juː]. This might indicate distinct pronunciations insisting more on some part of the W [ˈdʌbəljuː] or another.

Considering the maps obtained by the other DR methods (Fig. 6.19), **PCA**, **Isomap** and **NCA** obtain strongly overlapping classes, that may appear excessive based on the purity matrix. **S-Isomap** and **S-UMAP** on the other hand produce maps with over-separated classes, not representing the between class proximities

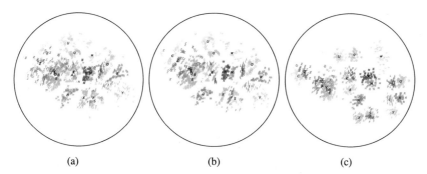

Fig. 6.18 ClassNeRV maps of the Isolet 5 dataset with $\tau^* = 0.5$ and different levels of supervision: $\epsilon = 0$ (no supervision), $\epsilon = 0.3$ (medium supervision) and $\epsilon = 0.5$ (maximum supervision). Classes centroids are represented by black markers. (**a**) No supervision (NeRV). (**b**) Medium supervision. (**c**) Maximum supervision

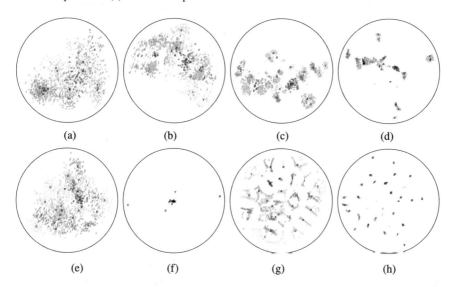

Fig. 6.19 Maps of the Isolet 5 dataset obtained by other DR techniques (both unsupervised and supervised). (**a**) PCA. (**b**) Isomap. (**c**) tSNE. (**d**) UMAP. (**e**) NCA. (**f**) S-Isomap. (**g**) Classimap. (**h**) S-UMAP

demonstrated by the purity matrix. Finally, UMAP and tSNE provide a relatively well-separated representation of classes, with overlaps between some of the classes mostly coherent with the purity matrix. Quality indicators presented Fig. 6.20 show trends similar to those observed for the digits dataset (Fig. 6.16), thus leading to the same conclusions.

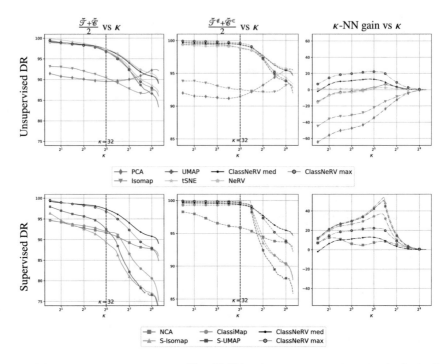

Fig. 6.20 Quantitative comparison of ClassNeRV and other DR techniques for the Isolet 5 dataset. Technicalities are similar to the caption of Fig. 6.13

6.6.5   Robustness to Class Misinformation

To study how ClassNeRV behaves in a case where the structure and classes are in conflict, we apply it to the digits dataset with random labels. That is, we attribute to each data point a label randomly selected in the integers from 0 to 9. Those labels are then fed to the algorithm. Figure 6.21 presents the maps of NeRV (with $\tau = 0.5$) and ClassNeRV (with maximum supervision) for that case. We may note that the position of points in NeRV map are the same as for the true labels (Fig. 6.14a), since NeRV is unsupervised.

For those maps, the shapes of the digits allow to determine whether data points mapped to close positions are effectively similar. Conversely, the colours indicate the random labels provided to the algorithm for each point. Since the classes and structure are unrelated, the classes should not be homogeneous in the map. Relying on the map, we may see that ClassNeRV mainly preserves the structure of data, with no perceivable increase of the class purity. This is supported by the indicators of Fig. 6.22. Indeed, the difference between ClassNeRV and NeRV is rather small for structure and class preservation, while values of the κ-NN gain remains low (under 5%). Note that, for the unsupervised methods, the structure preservation score $(\check{\mathcal{T}} + \check{\mathcal{C}})/2$ is strictly identical to the case with the true labels. As for the class preservation

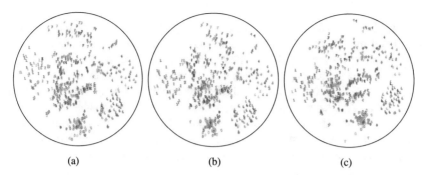

(a)   (b)   (c)

Fig. 6.21 ClassNeRV maps of the digits with random labels with $\tau^* = 0.5$ and different levels of supervision: $\epsilon = 0$ (no supervision), $\epsilon = 0.3$ (medium supervision) and $\epsilon = 0.5$ (maximum supervision).ClassNeRV digits (random labels). In the unsupervised case, the map is exactly the same as with the true labels. (**a**) No supervision (NeRV). (**b**) Medium supervision. (**c**) Maximum supervision

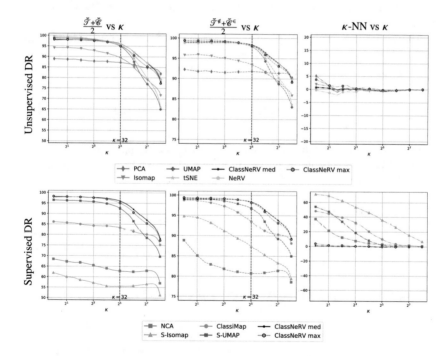

Fig. 6.22 Quantitative comparison of ClassNeRV and other DR techniques for the digits dataset (with random labels). Technicalities are similar to the caption of Fig. 6.13. Note that the class-aware indicators are computed with the random labels

score $(\check{\mathscr{T}}^{\notin} + \check{\mathscr{C}}^{\in})/2$, it accounts for a randomly selected subset of the distortions considered by the structure preservation indicator. Hence, it follows very similar trends. Finally, the κ-NN gain is almost null, since the class purity is close to $1/10$ both in the data space and in the map, with 10 the number of classes.

However, for the other supervised methods considered here, providing random labels induces a loss of structure preservation, as well as high values of the κ-NN gain. This implies that the randomly generated classes are strongly over-separated in the map, thereby distorting the structure. This indicates that ClassNeRV is more robust to mislabelled data than other supervised methods, since it privileges the structure over the classes.

6.6.6 Extension to the Type 2 Mixture: ClassJSE

The approach of ClassNeRV allows to introduce supervision in NeRV which uses a type 1 mixture of soft recall and soft precision. It may also be adapted to the supervision of JSE, which relies on a type 2 mixture of those terms. This adaptation yields another method called ClassJSE.

The stress of JSE introduced Sect. 5.3.3 is given by:

$$\zeta_i^{\text{JSE}} \triangleq \frac{1}{\tau}\mathscr{D}_{\text{KL}}(\beta_i, \mu_i) + \frac{1}{1-\tau}\mathscr{D}_{\text{KL}}(b_i, \mu_i), \tag{6.11}$$

where $\mu_i \triangleq (\mu_{ij})_{j\neq i}$, with $\mu_{ij} \triangleq \tau b_{ij} + (1-\tau)\beta_{ij}$. Applying a break-down similar to that of ClassNeRV given by Eq. (6.8), we obtain an expanded expression of ClassJSE stress:

$$\zeta_i^{\text{ClassJSE}} \triangleq \frac{1}{\tau^{\in}}\mathscr{D}_{\text{gKL}}(\beta_i^{\in}, \mu_i^{\in}) + \frac{1}{1-\tau^{\in}}\mathscr{D}_{\text{gKL}}(b_i^{\in}, \mu_i^{\in})$$
$$+ \frac{1}{\tau^{\notin}}\mathscr{D}_{\text{gKL}}(\beta_i^{\notin}, \mu_i^{\notin}) + \frac{1}{1-\tau^{\notin}}\mathscr{D}_{\text{gKL}}(b_i^{\notin}, \mu_i^{\notin}) \tag{6.12}$$

with the terms of distributions $\mu_i^{\in} \triangleq (\mu_{ij}^{\in})_{j\in\mathcal{S}_i^{\in}}$ and $\mu_i^{\notin} \triangleq (\mu_{ij}^{\notin})_{j\in\mathcal{S}_i^{\notin}}$ are defined as:

$$\mu_{ij}^{\in} \triangleq \tau^{\in}b_{ij} + (1-\tau^{\notin})\beta_{ij} \quad \text{and} \quad \mu_{ij}^{\notin} \triangleq \tau^{\notin}b_{ij} + (1-\tau^{\notin})\beta_{ij}.$$

In the unsupervised case ($\tau^{\in} = \tau^{\notin}$), grouping vertically the terms of Eq. (6.12) leads back to the stress of JSE given by Eq. (6.11).

Conversely, in the general case, the terms of Eq. (6.12) may be regrouped horizontally, leading to a more compact expression of ClassJSE stress:

$$\zeta_i^{\text{ClassJSE}} = \mathscr{M}_{\text{gKL},2}(\beta_i^{\in}, b_i^{\in}, \tau^{\in}) + \mathscr{M}_{\text{gKL},2}(\beta_i^{\notin}, b_i^{\notin}, \tau^{\notin}), \tag{6.13}$$

with the type 2 mixture between partial distributions $\beta_i^\in$ and $b_i^\in$:

$$\mathscr{M}_{\text{gKL},2}(\beta_i^\in, b_i^\in, \tau) \triangleq \frac{1}{\tau}\mathscr{D}_{\text{gKL}}(\beta_i^\in, \mu_i^\in) + \frac{1}{1-\tau}\mathscr{D}_{\text{gKL}}(b_i^\in, \mu_i^\in), \qquad (6.14)$$

with $\mu_{ij}^\in \triangleq \tau b_{ij} + (1-\tau)\beta_{ij}$.

The type 2 mixture of ClassJSE has several interesting properties. First, for any value of the trade-off parameter τ, the type 2 divergence between partial distributions $\mathscr{M}_{\text{gKL},2}(\beta_i^\in, b_i^\in, \tau)$ satisfies:

$$\mathscr{M}_{\text{gKL},2}(\beta_i^\in, b_i^\in, \tau) = \mathscr{M}_{\text{KL},2}(\beta_i^\in, b_i^\in, \tau),$$

where $\mathscr{M}_{\text{KL},2}(\beta_i^\in, b_i^\in, \tau)$ is defined by replacing $\mathscr{D}_{\text{gKL}}$ with $\mathscr{D}_{\text{KL}}$ in Eq. (6.14). This means that in JSE, the generalized Kullback–Leibler divergences may be replaced by classical Kullback–Leibler divergences without any impact.

In addition, the asymptotic behaviour of the type 2 mixture is as follows (see Annex C.2.2.3):

$$\mathscr{M}_{\text{gKL},2}(\beta_i^\in, b_i^\in, \tau) \xrightarrow[\tau\to 1]{} \mathscr{D}_{\text{gKL}}(\beta_i^\in, b_i^\in) \quad \text{and} \quad \mathscr{M}_{\text{gKL},2}(\beta_i^\in, b_i^\in, \tau) \xrightarrow[\tau\to 0]{} \mathscr{D}_{\text{gKL}}(b_i^\in, \beta_i^\in).$$

Thus, ClassJSE converges towards ClassNeRV when both trade-off parameters $\tau^\in$ and $\tau^{\neq}$ converge either towards 0 or 1. This result has already been shown in the unsupervised case, since JSE is known to converge towards NeRV for $\tau \to 0$ or $\tau \to 1$ [109]. We see here that ClassJSE also converges towards ClassNeRV in the most supervised case. Due to the closeness of formulation of ClassNeRV and ClassJSE, we denote the mixture by $\mathscr{M}_{\text{gKL},12}$ when both types of mixtures may be used equivalently.

Figures 6.23, 6.24 6.25, 6.26 show the sensitivity of ClassJSE to the trade-off parameters. The trends are similar to those observed with ClassNeRV. Indeed, considering the maps (Figs. 6.23 and 6.25), increasing the supervision steers the distortions to favour the separation of classes. For the sphere dataset, this leads to

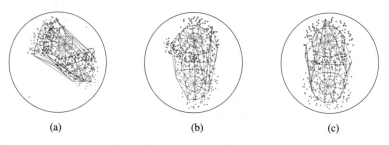

(a)   (b)   (c)

Fig. 6.23 ClassJSE maps of the sphere with $\tau^* = 0.5$ and different levels of supervision: $\epsilon = 0$ (no supervision), $\epsilon = 0.3$ (medium supervision) and $\epsilon = 0.5$ (maximum supervision). (**a**) No supervision (NeRV). (**b**) Medium supervision. (**c**) Maximum supervision

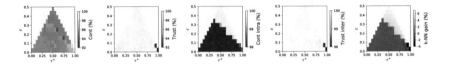

Fig. 6.24 Heatmaps of indicators for ClassJSE on the Sphere dataset in the τ^*, ϵ parameter space

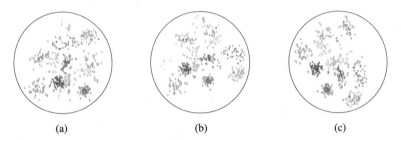

(a)   (b)   (c)

Fig. 6.25 ClassJSE maps of the digits with $\tau^* = 0.5$ and different levels of supervision: $\epsilon = 0$ (no supervision), $\epsilon = 0.3$ (medium supervision) and $\epsilon = 0.5$ (maximum supervision). (**a**) No supervision (NeRV). (**b**) Medium supervision. (**c**) Maximum supervision

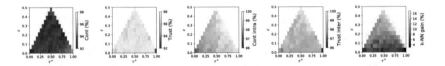

Fig. 6.26 Heatmaps of indicators for ClassJSE on the digits dataset in the τ^*, ϵ parameter space

a progressive displacement of the torn region towards the equator. Note that the map with maximum supervision is equivalent to that obtained with ClassNeRV, due to the common asymptotic behaviour of the two methods. We may note that the distortions that are visually noticed appear more severe, such as the two points strongly torn from the sphere in the unsupervised case. This may be explained based on the structure of the type 2 mixture. Indeed, conversely to the type 1 case, if $\beta_{ij} \gg 1$ and $b_{ij} \approx 0$ (severe missed neighbour), the log term does not explode, since μ_{ij} is not close to 0 (as an average of β_{ij} and b_{ij}). This may also be related to the plastic behaviour of JSE (detailed in Sect. 7.1.5.1). Due to this property, if a point is torn from its neighbourhood at some step of the optimization, there is no force bringing it back.

For the heatmaps of quality indicators (Figs. 6.24 and 6.26), we observe a general increase of false neighbours (worst trustworthiness values) and a decrease of missed neighbours (better continuity values) when τ^* increases, as well as a decrease of fault impacting classes when ϵ increases. Note that the bounds of the colour scales are the 5th and 95th percentile of the quality indicators, but for ClassNeRV maps.

6.6.7 Extension to Semi-Supervision and Weak-Supervision

In the above, Class-Guided Neighbourhood Embedding methods (i.e. ClassNeRV and ClassJSE) have been defined considering the fully supervised case, where each point belongs to a class. Inversely, in the semi-supervised case, some data points are not labelled. To handle those missing labels, all relations between points that both have labels may be treated exactly as in the fully supervised case, characterizing them either as intraclass or interclass. Yet, for pairs of points for which at least one have missing class-information a different strategy must be applied. This may be controlled by an independent trade-off parameter $\tau^\emptyset$ for the preservation of unsupervised relations. Hence, the stress of Class-Guided Neighbourhood Embedding methods may be redefined as follows to handle missing labels:

$$\zeta_i^{\text{CGNE}} = \mathcal{M}_{\text{gKL},12}(b_i^{\in}, \beta_i^{\in}, \tau^{\in}) + \mathcal{M}_{\text{gKL},12}(b_i^{\notin}, \beta_i^{\notin}, \tau^{\notin}) + \mathcal{M}_{\text{gKL},12}(b_i^{\emptyset}, \beta_i^{\emptyset}, \tau^{\emptyset}), \tag{6.15}$$

where $\beta_i^{\emptyset} \triangleq (\beta_{ij})_{j \in S_i^{\emptyset}}$, with a symmetric definition for $b_i^{\emptyset}$. In this approach, the sets $S_i^{\in}$, $S_i^{\notin}$ and $S_i^{\emptyset}$ respectively listing intraclass, interclass and uncharacterised neighbours, are redefined as follows:

$$S_i^{\emptyset} \triangleq \{j \neq i \mid L_i = -1 \cup L_j = -1\},$$
$$S_i^{\in} \triangleq \{j \neq i \mid L_j = L_i\} \setminus S_i^{\emptyset},$$
$$S_i^{\notin} \triangleq \{j \neq i \mid L_j \neq L_i\} \setminus S_i^{\emptyset},$$

where $L_i = -1$ denotes a missing label for the point i.

The choice of trade-off parameters should respect $\tau^{\in} \geqslant \tau^{\emptyset} \geqslant \tau^{\notin}$, ensuring for the uncharacterised relations an intermediate behaviour between those of intraclass and interclass relations. A default parametrization could be $\tau^{\emptyset} = \tau^*$. Yet, this parameter may be adjustednto decide whether uncharacterised points should be treated more as intraclass or interclass neighbours.

The formulation of Eq. (6.15) also extends the framework to the weakly supervised case with pairwise information [15]. In that case the supervision only provides sets of points pairs that should be in the same class or that should be different. Those sets may be used directly to define intraclass and interclass neighbours $S_i^{\in}$ and $S_i^{\notin}$ for a given point i, while pairs without class-information define the uncharacterized neighbours $S_i^{\emptyset}$. This weakly supervised approach could be coupled with visual interaction, allowing users to define "must be" and "must not be" similarities, in order to steer the embedding.

Figure 6.27 shows the result of ClassNeRV for the digits dataset, with one eighth of the labels removed. For NeRV map, the position are not affected, and are identical to the previous cases. However, for ClassNeRV, the classes appear less homogeneous than in the fully supervised case. For example, with maximum

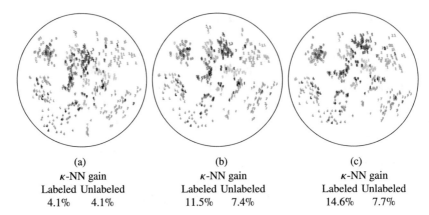

Fig. 6.27 ClassNeRV maps of the digits with missing labels with $\tau^* = \tau^{\emptyset} = 0.5$ and different levels of supervision: $\epsilon = 0$ (no supervision), $\epsilon = 0.3$ (medium supervision) and $\epsilon = 0.5$ (maximum supervision). In the unsupervised case, the map is exactly the same as with the true labels. Digits with missing labels are represented in black. κ-NN gain is added for $\kappa = 32$, aggregating it separately for the labelled and unlabelled data points. That indicator is computed with the true labels. (**a**) No supervision (NeRV). (**b**) Medium supervision. (**c**) Maximum supervision

supervision, the 9s were represented as one group in the fully supervised case (Fig. 6.14c), while they are separated into two groups in the semi-supervised case (Fig. 6.27c). This may come from a lack of class-information along the torn frontier, which would allow more missed neighbours. We see however, that the unlabelled points are represented close from similar points. To quantify how the unlabelled points are regrouped with their true class, we compute for each map the average κ-NN gain for the labelled and unlabelled points, as presented Fig. 6.27. It shows that the class purity increases with the level of supervision for the labelled points, but also, to a lesser degree, for the unlabelled points. Hence, the global increase of class separation also induces a stronger grouping of unlabelled points with elements of their class.

6.6.8   Extension to Soft Labels

Hard labels $\{L_i\}$ indicate with absolute certainty the class to which the point i belongs. Yet, the class-information may contain uncertainty leading to the use of soft-labels $\{\tilde{L}_i\}$. Those soft-labels correspond to a distribution of probabilities that a point belongs to one class or another. In that case, the relations may not be discriminated into within-class and between-class relations. Instead, we may define a class-community measure $q_{ij} \in [0, 1]$ quantifying the similarity of the soft labels $\tilde{L}_i$ and $\tilde{L}_j$. where 0 correspond to no class-community and 1 to maximum class-community. Thus, a different trade-off parameter τ_{ij} may be defined for each pair of points (i, j), based on their class-community. A simple solution would be:

$$\tau_{ij} = q_{ij}\tau^{\in} + (1 - q_{ij})\tau^{\notin}.$$

This leads to reformulate CGNE stress with the following general formulation:

$$\zeta_i^{\mathrm{CGNE}} = \mathscr{M}_{\mathrm{gKL},12}(b_i, \beta_i, \tau_i).$$

In that expression $\mathscr{M}_{\mathrm{gKL},12}(b_i, \beta_i, \tau_i)$ may correspond to the type 1 mixture (for ClassNeRV):

$$\mathscr{M}_{\mathrm{gKL},1}(b_i, \beta_i, \tau_i) = \sum_{j \neq i} \tau_{ij} \mathscr{D}_{\mathrm{B}}(\beta_{ij}, b_{ij}) + (1 - \tau_{ij})\mathscr{D}_{\mathrm{B}}(b_{ij}, \beta_{ij})$$

or to the type 2 mixture (for ClassJSE):

$$\mathscr{M}_{\mathrm{gKL},2}(b_i, \beta_i, \tau_i) = \sum_{j \neq i} \frac{1}{\tau_{ij}} \mathscr{D}_{\mathrm{B}}(\beta_{ij}, \mu_{ij}) + \frac{1}{1 - \tau_{ij}} \mathscr{D}_{\mathrm{B}}(b_{ij}, \mu_{ij})$$

where $\mu_{ij} = \tau_{ij}b_{ij} + (1 - \tau_{ij})\beta_{ij}$, and the scalar Bregman divergence between terms β_{ij} and b_{ij} is given by

$$\mathscr{D}_{\mathrm{B}}(\beta_{ij}, b_{ij}) = \beta_{ij}\log\left(\frac{\beta_{ij}}{b_{ij}}\right) + b_{ij} - \beta_{ij}.$$

This formulation leads back to the case with hard labels $\{L_i\}$ using the following definition of class community:

$$q_{ij} = \begin{cases} q_{ij} = 1_{L_i - L_j} & \text{if } L_i \neq -1 \text{ and } I_{.j} \neq -1, \\ q_{ij} = 0.5 & \text{if } L_i = -1 \text{ or } L_j = -1. \end{cases}$$

Note that this class-community approach could extend to multi-labels or hierarchical class-structures.

6.6.9 Intermediate Conclusions

CGNE methods allows to benefit from class-information to obtain a low-dimensional map of metric data ensuring both class cohesion and class distinction. Its level of supervision may be tuned to focus more on the preservation of the data structure or on the separation of classes. As such, it provides an interesting solution for visual exploration of labelled data. This can be used to get insights on the structure of classes with respect to one another, but also to assess the discriminability of classes in a classification approach. The latter could lead to question the considered data features or to verify the labels affected to each data instance. Finally, CGNE methods can easily be adapted to incomplete, weak or uncertain class-information.

Future work could extend that framework to the supervision of other methods such as tSNE or UMAP, for which there is no such symmetric effects penalizing false and missed neighbours. Another research perspective could be to build a criterion to set automatically the best hyper-parameters providing an ideal trade-off between the several adversarial objectives of the methods.

Chapter 7
Optimization, Acceleration and Out of Sample Extensions

> *The effective numerical treatment of partial differential equations is not a handicraft, but an art.*
>
> Folklore

This chapter details the challenges involved for the optimization of the stress functions presented in the previous chapters, as well as the solutions effectively used (Sect. 7.1). It then addresses the issue of time and space complexity and DR techniques acceleration (Sect. 7.2). Finally, it dives into out of sample extension of the mapping, which allows to position new data points a posteriori onto a given map (Sect. 7.3).

7.1 Optimization

The vast majority of DR methods rely on the minimization of a stress function $\zeta : \mathcal{E} \longrightarrow \mathbb{R}$, also called objective, cost, loss or energy function. This optimization process consists in searching in the solution space $\mathcal{E}$ the values of the problem variables leading to the minimum value of the considered stress function. For non-parametric DR methods [54, 87, 109, 125, 158, 183, 189], the stress is directly optimized with respect to the coordinates of the embedded points X, leading to:

$$X^* = \operatorname*{argmin}_{X \in \mathcal{E}} \zeta(X).$$

Conversely, for parametric methods [67, 79, 88, 141, 173], the stress is minimized with respect to a parameters vector W of a parametric mapping Φ_W, so that:

$$X^* = \Phi_{W^*}(\Xi) \quad \text{with} \quad W^* = \operatorname*{argmin}_{W \in \mathcal{W}} \zeta\left(\Phi_W(\Xi)\right).$$

Hence, in the parametric case, the solution space is a parameter space $\mathcal{W}$ and not the embedding space $\mathcal{E}$. Due to optimization challenges, the optimal embedding X^*

S. Lespinats et al., *Nonlinear Dimensionality Reduction Techniques*,
https://doi.org/10.1007/978-3-030-81026-9_7

with respect to ζ is often not obtained in practice, and the DR method can provide a suboptimal solution (local optimum). In the following, we will focus on the non-parametric case of optimization.

7.1.1   Global and Local Optima

One of the key difficulties of optimization lies in the difference between local and global optima. Considering a minimization problem, a *local optimum* is a solution X^* minimizing the stress ζ over a neighbourhood of the solution space, i.e. such that $\zeta(X^*) \leqslant \zeta(X)$ for all X in some neighbourhood of X^*. The local optimum X^* is also a *global optimum* if it minimizes the stress ζ for the entire solution space. The ultimate goal of optimization algorithms is to find a global minimum of the stress function. Yet, state-of-the-art optimization algorithms, such as gradient descent, are only designed to converge to a local optimum. Since it is not feasible to explore the entire solution space (except for small discrete optimization problems), it remains impossible to determine whether it is in fact a global optimum, except for very specific cases. Those exceptions include the convex functions optimized over a convex solution space, for which any local optimum is guaranteed to be a global optimum. This includes the case of spectral projection techniques (see Sect. 5.1).

7.1.2   Gradient Descent and Quasi-Newton Methods

Descent algorithms, such as gradient descent or quasi-Newton methods, rely on local properties of the stress function to orientate their exploration of the solution space. At iteration t, they attempt to improve the current solution $X^{(t)}$ by moving along a descent direction $p^{(t)}$. That descent iteration may be simply expressed as:

$$X^{(t+1)} = X^{(t)} + \eta^{(t)} p^{(t)},$$

with $\eta^{(t)}$ the step size (or learning rate). The result of those algorithms strongly depends on the methods used to define the direction $p^{(t)}$, step size $\eta^{(t)}$ and initialization $X^{(0)}$. Figure 7.1 illustrates some tendencies involved in that choice for the simple case of quadratic stress functions.

The gradient descent (or steepest descent) algorithm relies on first order local approximation of the function ζ. Thus, it defines the descent direction as the opposite of the gradient:

$$p^{(t)} = -\nabla \zeta \left(X^{(t)} \right),$$

with $\nabla \zeta$ the gradient of the stress.

Quasi-Newton methods [138], such as the Broyden–Fletcher–Goldfarb–Shanno (BFGS) and Low-memory BFGS (L–BFGS), incorporate additional information

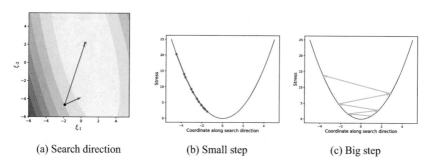

(a) Search direction   (b) Small step   (c) Big step

Fig. 7.1 Effect of descent direction and step size for the minimization of hypothetical quadratic functions with gradient-based methods. Panel (**a**) shows steps of a gradient descent (blue arrow) and Newton method (red arrow) in the two-dimensional solution space. Panels (**b**) and (**c**) illustrate the impact of the step size considering a fixed search direction. A too small step size may require many iterations to converge, whereas a too big step size can lead to oscillations around the minimum or even divergence

about the curvature, relying on a second order approximation of ζ (i.e. considering it as locally quadratic). The BFGS algorithm is used for minimizing the stress of **CGNE** methods (Sect. 6.6) and **ASKI** (see Sect. 5.3.7). Intuitively, this means that the steepest descent direction is corrected considering that the minimum is closer in directions of strong curvature than in directions of mild curvature. The descent direction is thus given by:

$$p^{(t)} = -\widehat{\mathbf{H}}_\zeta (X_t)^{-1} \nabla \zeta \left(X^{(t)} \right)$$

with $\widehat{\mathbf{H}}_\zeta$ an approximation of the Hessian of the stress function obtained from previous values of the gradient. Indeed, computing the exact Hessian may be computationally costly. This approach amount to using the Newton root-finding method to locate the zero of the gradient (which indicates a local optimum). Note that other alternatives are employed around these main concepts (e.g. conjugate gradient [189], simulated annealing [87]).

Figure 7.1a illustrates the descent direction found by gradient descent (blue arrow) and exact Newton method (red arrow). Once $p^{(t)}$ is determined, the optimization problem becomes 1−dimensional and amounts to finding the step size minimizing the stress along that direction. Such minimization can rely on line-search strategies [138], or on heuristic definitions of the step size [183]. For the latter case, there is a trade-off between a two small step size, which could lead to very slow convergence (see Fig. 7.1b), or a too big step size which could lead to divergence (oscillating around the optimum). A possible strategy is to use a step size progressively decreasing with the number of iteration, so that the algorithm get faster near the optimum and then avoid divergence [125, 209]. In the case of quasi-Newton methods, a natural value of the step size is $\eta^{(t)} = 1$ [138]. For the quadratic case shown Fig. 7.1a, this leads to the minimum in only one iteration (as shown by the red arrow).

7.1.3   Initialization

The initial map is either obtained by random positioning of the points [183, 187, 189], or using spectral methods [110, 125, 189]. The interest of spectral methods, is that their stress is convex, such that the global optimum is always reached. Though these methods do not provide an ideal representation in terms of the specific non-convex stresses, they provide a good guess of some global map structure. CGNE and ASKI are first initialized with cMDS [177]. This method is equivalent to PCA in a Euclidean space, but is adapted to any metric space since it relies only on a distance matrix.

7.1.4   Multi-Scale Optimization

Many DR methods incorporate a scale parameter (e.g., radius, number of neighbours or perplexity) in their stress function, defining the size of the neighbourhoods that they attempt to preserve. The values of such parameters are usually set to retain local information at a relatively small scale. A multi-scale optimization strategy [110, 187, 189] consists in progressively decreasing that scale parameter during the optimization process down to a target value. Intuitively, this strategy allows to first obtain a good representation of the global structure, coarsely regrouping the points, and, then, to refine the details by improving the local structure (which is the final focus). Conversely, in the case of a single-scale optimization, the structure is only rearranged very locally and may be subject to more local optima.

Figure 7.2 illustrates this idea with the stress function of NeRV, for a 1−dimensional embedding with all but one point perfectly positioned. For a high scale (Fig. 7.2a), the stress is relatively smooth, allowing to converge towards the global optimum of that simple case. Yet, when the scale decreases (Fig. 7.2b and c), the stress becomes subject to variations of higher spatial frequency. This renders the final solution more precise, with a clearer minimum (reducing the size of the plateau around the minimum). Yet, it increases the probability that the optimization remains trapped in a local optimum. We may note that, for a given configuration of the embedded points, the value of the stress of NeRV tend to increase when the scale decreases. Hence, during a multi-scale optimization, the value of the stress may increase between successive iterations.

Multi-scale optimization for neighbourhood embedding methods either decreases linearly the scale parameter σ_i, from $\Delta_{\max}/2$ to the target value [189], or the entropy $H(\sigma_i)$ [110] (as defined in Sect. 5.3). In the latter case, the perplexity decreases from $N/2$ to the target value with a geometric progression of ratio $1/2$.

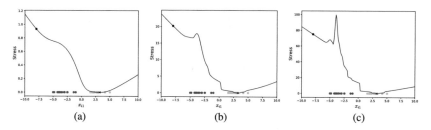

Fig. 7.2 Stress function for an intrinsically one-dimensional dataset embedded in a one-dimensional space as a function of the position of one point, the others (blue and orange) being placed at their optimum position. High scales lead to lower frequency variations of the stress reducing the risk of getting trapped in a local optimum, whereas low scales give a sharper valley around the precise global optimum (black cross). A possible initialization of the variable point is represented by the black circle. (**a**) $\kappa^\star = 16$. (**b**) $\kappa^\star = 8$. (**c**) $\kappa^\star = 4$

This second approach is used for ASKI and CGNE, along with an initializaiton with cMDS.

7.1.5 Force-Directed Placement Interpretation

Force-directed placement [70, 131] proposes to interpret the optimization of DR stress functions using gradient descent based on a mechanical analogy. In that analogy, the optimization process searches a stable state of an hypothetical mechanical system constituted of the embedded points with each pair of points being connected by a spring. Indeed, the gradient of DR stress with respect to the embedded points coordinates (assuming Euclidean distances in the embedding space) is of the form (see Annex C.1):

$$-\frac{\partial \zeta}{\partial x_i} = \sum_{j \neq i} F_{ij} \frac{x_j - x_i}{D_{ij}},$$

with $F_{ij} = \frac{\partial \zeta}{\partial D_{ij}} + \frac{\partial \zeta}{\partial D_{ji}}$. Thus, each point j exerts on i a force of value F_{ij} and directed from x_i to x_j. A positive value corresponds to an attractive force and a negative value to a repulsive force.

7.1.5.1 Elastic and Plastic Behaviours

This model may be related to elastic and plastic mechanical behaviours in materials [107]. Indeed, before reaching its elastic limit a strained spring exerts a force that tends to bring it back to its equilibrium length. In the analogy, the equilibrium length of the spring between points i and j is the data space distance Δ_{ij} (assuming a

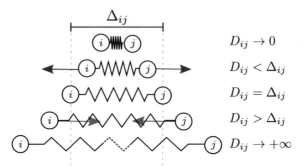

Fig. 7.3 Elastic and plastic behaviours of springs for an hypothetical metric DR method with pairwise separable stress (i.e. with minimum reached for $\Delta_{ij} = D_{ij}$). From top to bottom: plastic compression (no force), elastic compression (repulsive force), spring at its equilibrium length (no force), elastic elongation (attractive force) and plastic elongation (no force)

metric technique with pairwise separability of the stress). For a well-represented distance ($D_{ij} = \Delta_{ij}$) the spring exerts no force. Inversely, in case of a distortion of the relation $i \sim j$, a compressed spring ($D_{ij} < \Delta_{ij}$) will exert a repulsive force, while a stretched spring ($D_{ij} > \Delta_{ij}$) will exert an attractive force. Yet, in the case of plastic deformations, these forces may vanish for excessive distortions of the relation. In that case the springs may be subject to plastic compressions (if $\lim_{D_{ij} \to 0} F_{ij} = 0$) or plastic elongations (if $\lim_{D_{ij} \to 0} F_{ij} = +\infty$). Those different behaviours are illustrated Fig. 7.3. Note that for non-metric and/or non-separable stress functions, the expression of the equilibrium length of the spring between i and j may be more complex. Indeed, it may also account for a scaling factor or depend on the position of other neighbouring points in the map, as for neighbourhood embedding methods.

The elastic or plastic behaviour of a technique influences the result reached by the optimization process, especially for non-mappable data (necessarily sub-optimal embedding). Indeed, in the elastic case, a sub-optimal embedding will consist in an (unstable) equilibrium state with attractive and repulsive forces cancelling each other. Each of those forces attempts to return one of the springs to its ideal size. Inversely, in the plastic case, such an equilibrium state may be attained by "breaking" certain springs. This amounts to loosening some of the constraints, admitting that some false neighbours will remain close (plastic compressions) or that some missed neighbours will never join each other (plastic elongation).

7.1.5.2 Stochastic Gradient Descent

In the framework of force-directed placement, stochastic gradient descent approaches [54, 125, 187] may be seen as decompressing one spring at a time in a random order (reshuffling at each epoch) [209]. This interpretation allows to provide an upper bound of the step size to avoid oscillations around the equilibrium length of the spring.

7.1.5.3 Attractive-Repulsive Decomposition

The forces F_{ij} may also be considered through a decomposition into a positive part F_{ij}^+ and a negative part F_{ij}^- [31, 193]. Depending on the respective cumulated magnitudes of the attractive components F_{ij}^+ and of the repulsive components F_{ij}^-, DR methods may be ordered along an attraction-repulsion spectrum [20]. In terms of structure distortions, more attractive methods tend to induce more manifold compressions, whereas more repulsive lead to more manifold stretching.

The optimization trick of early exaggeration used for the optimization of tSNE stress [183] may be interpreted in that optic. Indeed, it artificially increases the attractive forces in the early phase of the embedding process (multiplying data space membership degrees by a constant higher than 1) to avoid irreversible plastic elongation at that stage.

For stress function focusing on the preservation of neighbourhood relations at a small scale, the attractive component F_{ij}^+ is higher for pairs of points (i, j) that are neighbours in the data space, while the repulsive component F_{ij}^- is higher for those that are neighbours in the embedding space. Thus, for reliable neighbours those components tend to cancel out while for non-neighbours, they both are negligible. Conversely, false neighbours tend to exert short-range repulsive forces on each other, and missed neighbours exert long range attractive forces. This difference of range comes from the fact that false neighbours are those that are close in the embedding space by definition.

7.1.5.4 Blockade Effect

In a low dimensional embedding space, this combination of attractive and repulsive forces may lead to poor local optima [193], as illustrated by Fig. 7.4. In that Figure, the black point is attracted towards its missed neighbours but also repulsed by false neighbours placed in-between. Moreover, due to the difference of range of

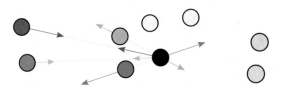

Fig. 7.4 Blockade effect pressuring the black point into a local optimum. Arrows represent forces involving the black point. Missed neighbours (in shades or red) exert long-range attractive forces, while false neighbours (in shades of blue) exert short-range repulsive forces. Reliable neighbours (pale yellow) and non-neighbours (grey) exert negligible forces, considering that the magnitude of the force increases with distortion severity

those forces, the magnitude of repulsing forces increases when the black points get closer from its false neighbours. Hence, at some point, the forces cancel out, thus maintaining the black point in a sub-optimal equilibrium statey cancelling out leading to a sub-optimal equilibrium. This phenomenon may also be interpreted in terms of stress function considering Fig. 7.2b. In that case, the black point is attracted by its missed neighbours (orange cluster) and repulsed by its false neighbours (blue cluster), leading to a local optimum on the wrong side of the blue cluster. As such, the blue cluster placed in-between the point and its ideal position constitutes a blockade, that may not be trespassed.

7.1.5.5   Auxiliary Dimensions

A solution to such a problem can be to provide additional degrees of freedom allowing the point to get around the blockade. In terms of stress, this might be interpreted as converting a local optimum into a saddle point. This may be obtained by embedding the points in a space of dimensionality higher than the target value and then progressively penalizing the coordinates of points along the additional dimensions [87]. With that approach, all points benefit from a wider range of trajectories to move around the space in the early phase of optimization.

Another solution first obtains an initial map and identifies points potentially subject to the blockade effect, named pressured points [193]. Pressured points are those that would move the most from their current position if they where provided an additional dimension (considering an optimization only along that new coordinate). The optimization of all points positions is then continued providing an additional dimension only a restricted set of pressured points (updated at each iteration). In that regard, the pressure point approach uses some sort of local map evaluation (the level of pressure indicating that the points are subject to much compression), to steer new iterations of the DR process and improve the map.

7.2   Acceleration Strategies

The *time complexity* of an algorithm corresponds to an estimation of the number of operations necessary to reach the result. It relates to a time based on the number of floating point operations (flops) that the machine can perform during 1 s. The *space complexity* corresponds to the maximum amount of memory (number of bytes) required by the algorithm to store its internal variables.

Thus, for DR techniques operating on full matrices of pairwise similarities or dissimilarities, the time and space complexity is at least $O(N^2)$ for each iteration, with N the number of data points. In order to enable the analysis of massive datasets ($N \gg 1$) as well as more interactive use of DR algorithms (low mapping time), it is necessary to lower that complexity through acceleration strategies.

Considering a gradient descent optimization with an empirical definition of the step size, only the computation of the gradient is required. In the force-directed placement approach, the computation of that gradient requires to estimate $O(N^2)$ forces between each pair of points. For local methods, forces are predominant between points that are neighbours in the data or embedding space. Hence, the complexity may be lowered by approximating the gradient computing a reduced number of forces.

7.2.1 Attractive Forces Approximation

Attractive forces ($F_{ij} > 0$) are predominant between pairs of points that are close in the data space. Those data space neighbourhoods are constant all along the optimization process. Furthermore, for local methods defining the neighbourhood size based on a fixed number of neighbours or perplexity value, the number of pairs is ensured to be small. Hence, the computation of the gradient may be accelerated by neglecting attractive forces between points that are not neighbours in the data space. For graph layout techniques embedding the κ-NN graph, the attractive forces are only considered between the κ nearest neighbours [125, 174], while for tSNE acceleration [182], they are limited to the $3\kappa^\star$ neighbourhood, with $\kappa^\star$ the perplexity. Hence, the computation of attractive forces has a complexity of $O(N\kappa)$, which is almost linear for $\kappa \ll N$.

Yet, this approximation strategy requires to be able to identify the nearest neighbours with a relatively low complexity. A brute force approach necessitates to compute the complete distance matrix, finding the κ smallest values of its row, leading to a complexity at least quadratic, that is $O(N^2)$ [56]. A more subtle approach, consists in using binary search trees, namely some sort of indexing of multidimensional spaces.

7.2.2 Binary Search Trees

A binary search tree [45, 69, 180, 205] is a tree representing a hierarchical partition of a space. Each node of the tree corresponds to a region of that space called a cell, the root containing the entire space. Each non-leaf node also defines a decision rule splitting the cell into two sub-cells associated with the two children of the node. The tree is first built considering a given set of points, usually so that, at each level of the tree, the cells all contain an approximately balanced number of points. This implies that each decision rule approximately splits its parent cell into two roughly balanced cells. We may note that a perfectly balanced tree splitting points into leaf cells each containing one point requires $\lceil \log_2(N) \rceil$ levels. In practice leaf cells contain more points, hence requiring less levels.

Once the tree is built, any new point (or query) may be attributed to a leaf cell by successively testing each decision rule. Considering that the diameter of a cell (i.e. the distance between the most distant pair of points in the cell) quickly decreases at each level of the tree (traversing towards the leaves), the points in the same cell as the query are likely to be its nearest neighbours. Thus, a balanced tree meeting this condition of cell diameter decrease enables to find the nearest neighbours of N points with a complexity of $O(N \log N)$.

The k-d trees [69] partition a Euclidean space in hyper-rectangular cells. Each decision rule splits the cell along a direction defined by one of the variables, placing the limit at a given threshold value. This direction is chosen as the one inducing the most spread for the cell points and the threshold can be set as the median of the variable for those points to obtain a balanced tree. This type of trees is very efficient for low dimensional Euclidean spaces. Yet, it is not robust for high dimensional spaces or general metric spaces.

Metric trees (also called vantage-point trees) [180, 205] are designed for metric spaces of any dimensionality. For node of a metric tree, the decision rule splits points depending on the distance from a vantage point, dividing the space between inside and outside a ball centred at that point. The distance threshold, namely the radius of that ball, may be taken as the median of the distances of the cell points to the vantage point. A good tree requires well-distributed vantage points. Metric trees are used in [182].

Random projection trees [45] implement a strategy close from the k-d tree, but are better-suited for points sampled from a low dimensional manifold in an ambient Euclidean space of high dimensionality. Each cell split is performed along a hyperplane orthogonal to a direction randomly sampled from the unit hypersphere. Hence, the decision rule divide points based on their projection on that random axis. The threshold value is obtained by taking the median and adding a random noise with specific properties. For those conditions, it may be shown that $\partial \log \partial$ levels are sufficient to at least halve the diameter of a cell, where ∂ is the fractal dimensionality estimated with Assouad method [45]. Random projection trees may be combined with nearest neighbours descent [56] as proposed by Tang et al. [174].

7.2.3 Repulsive Forces

Repulsive forces ($F_{ij} < 0$) can be significant between all point pairs that are close in the embedding space. During the optimization process those pairs may change, preventing to determine them once and for all. In addition, the number of points in the embedding space neighbourhood is often unconstrained, since the neighbourhood size is mostly defined by a fixed radius, rather than a number of neighbours or perplexity. Finally, the decrease of forces magnitude with distance may be slower than in the data space, for instance when point similarities are measured with heavy-tailed kernel. In that case, forces may not be strictly neglected.

As such, two main strategies may be used to reduce the $O(N^2)$ complexity of repulsive forces computation: forces sampling and forces aggregation.

7.2.3.1   Forces Sampling

For a separable stress of the form $\zeta = \sum_i \sum_{j \neq i} \zeta_{ij}$ where the pairwise stress ζ_{ij} only depends on its associated distance D_{ij} (and not on other map distances) the forces are independent of each other. In that case, optimization may be accelerated relying on the Stochastic Gradient Descent (SGD). At each iteration of the descent, the negative sampling approach [125, 174] considers all attractive forces exerted on point i by its data space neighbours, but only a random sample of the repulsive forces exerted on i. Moreover, the force between i and j is sampled with a probability proportional to the degree of vertex j in the κ-NN graph of the data space. With negative sampling the complexity of gradient computation becomes $O(Nm)$ with m the number of sampled repulsive forces at each iteration of the SGD, which is near linear when $m \ll N$. However, this strategy modifies the global equilibrium between attractive and repulsive forces compared with the exact gradient [20].

7.2.3.2   Forces Aggregation

For the case of non-separable stress functions, each force depends on all distances, so that all points should be accounted for at each iteration. Hence, in that case gradient approximation is performed by aggregating repulsive forces. Such acceleration strategies have been proposed for reducing the complexity of tSNE [147, 182]. In the gradient of tSNE, forces between each pair of points (i, j) may be separated into a positive part F_{ij}^+ (always attractive) and a negative part F_{ij}^- (always repulsive). The attractive part may be approximated as detailed in Sect. 7.2.1, while the negative part only depends on the respective position of points i and j in the map and on a normalizing term $Z = \sum_i \sum_{j \neq i} (1 + D_{ij}^2)^{-1}$.

Barnes–Hut approximation of tSNE [182] relies on an approximation used for the simulation of N-body problems in mechanics. In that approach, the sum of forces exerted on a point i by a group of n points whose diameter is small compared with its distance to i is approximated by n times the force exerted on i by the centroid of that group. In practice, Barnes-Hut tSNE builds at each iteration of the gradient descent a k-d tree in the map. Then, for each point i the cells that may be aggregated as one point are identified. This is obtained by performing a depth-first traversal of the k-d tree, successively splitting each cell until it is sufficiently small. This stopping criterion is reached when the ratio between the cell diameter and the distance between i and the cell centroid becomes smaller than a given threshold. The complexity of this approach is in $O(N \log N)$ and increases when the threshold increases. For very high values of that threshold, Barnes-Hutt method compute all forces individually leading back to the exact solution. A rather similar approach is proposed by Yang et al. [203] in the general case of Neighbourhood Embedding methods.

Another possibility for aggregating tSNE repulsive forces relies on texture splatting with a Graphical Processing Units (GPU). In that configuration, a scalar field over the embedding space may be stored as a texture of p pixels (assuming an embedding dimensionality of $d = 2$). Hence, the vector field of repulsive forces may be given by two textures. These fields are assessed by summing the fields induced by each individual point i using texture splatting. Thus, the sum of forces may be computed in $O(Np)$ computations with a memory usage of $O(p)$. Then, the cumulated forces are successively applied to all points i by texture interpolation, requiring $O(Np)$ computations. Note that due to the GPU architecture, the $O(p)$ computations for texture splatting or texture interpolation are ran in parallel, leading to an effective time complexity of $O(N)$.

In both cases, the normalization term Z is also obtained with the k-d tree or using a point density texture, thus not impacting the complexity of those approximations. However, the efficiency of those force aggregation approaches significantly drops when the embedding dimensionality increases.

7.2.4 Landmarks Approximation

The aforementioned acceleration approaches only work for a few very local methods with a scale parameter $\kappa \ll N$. Yet, in other cases, the computational complexity of a DR technique embedding N data points remains $O(N^2)$ or higher. Hence, a possible acceleration strategy is to map only a sub-sample of $n \ll N$ landmarks chosen among the data points, thus reducing the complexity to $O(n^2)$. The remaining data points may then be mapped relying on out-of-sample extension of that initial mapping, as detailed in the next Section.

7.3 Out of Sample Extension

Dimensionality reduction techniques allow to define a mapping Φ linking a discrete set of data points $\{\xi_i\}$ to corresponding embedded points $\{x_i\}$, with $x_i = \Phi(\xi_i)$. This mapping may either be parametric or non-parametric. A parametric mapping relies on an underlying function $\widehat{\Phi} : \mathcal{D} \longrightarrow \mathcal{E}$ defined on the entire data space, where Φ is the restriction of $\widehat{\Phi}$ to the set $\{\xi_i\}$. Conversely, a non-parametric mapping is only defined for the data points. In both cases, the mapping Φ resulting from the application of a DR technique depends on the data samples that are considered. Thus, if an already embedded dataset $\{\xi_i\}$ is extended by adding new data points $\{\xi_i'\}$, embedding that extended dataset $\{\xi_i\} \cup \{\xi_i'\}$ by applying a DR technique would require to start over the embedding process. Yet, this is computationally expensive and may lead to very different embedding space positions for the initial sample. Another solution would be to extend the pre-existing mapping to the new points using a posteriori positioning methods (also called out of sample extension).

7.3.1 Applications

Out of sample extension may be useful in different context, comprising mapping acceleration, mapping interaction, incremental positioning and classification in the map.

7.3.1.1 Mapping Acceleration

The computational cost of DR techniques tend to explode with the number of data points to be embedded, with a complexity of at least $O(N^2)$ at each gradient iteration for methods accounting for all pairwise relations. To lower that complexity, out of sample extension may be used [16, 52, 140, 143]. This involves the following steps. First, a set of $n \ll N$ landmarks (or pivots) is selected, that should be representative of the overall dataset. Then, an initial mapping is computed for these points in a rather short time (e.g., $O(n^2)$ for DR techniques with quadratic complexity). Finally, this mapping is extended to all the other points. Assuming that each evaluation of the extended mapping $\widehat{\Phi}$ has a complexity of $O(n)$, this last phase requires $O(Nn)$ computations for positioning the entire dataset.

7.3.1.2 Mapping Interaction

A posteriori positioning may also be used to enable interaction between the user and the mapping [5, 93]. In this framework, the initial mapping Φ is defined for a set of n control points, and extended to the complete dataset through out of sample extension. Hence, the analyst may interact with the mapping by simply moving a few control points. At each change, the non-control points are re-positioned in near-real time, thus adapting the entire map. This may for instance be used to incorporate user knowledge in the mapping [5]. To facilitate interaction, the number of control points n must also remain small.

7.3.1.3 Incremental Positioning: Data Streams

Data streams correspond to a specific case where the dataset changes over time, with new data points added continuously. In that case, the mapping process is subject to two key challenges: computational cost and cognitive load [71]. The computational cost comes from the algorithmic computations needed to update the projection at each iteration in order to incorporate the new points. The cognitive load affects the analyst if the positions of all points in the mapping change drastically at each step. To address those issues, [71] relies on a specifically designed incremental DR technique extending **PCA**, and on Procrustes transform allowing to preserve as well as possible the positions of points between successive maps. In the general case, considering no assumption on the DR technique, a posteriori positioning could be a good solution to both those problems. Indeed, it allows to position fast a new point based on a pre-existing mapping without changing the previous position of points. Yet, if the addition of new points significantly changes the structure of the dataset, that approach should be combined with regular map refinement by optimization of the DR stress for the entire dataset.

7.3.1.4 Classification in the Map

Out of sample extension may also enable to perform classification in the embedding space, for instance relying on a k-NN classifier [75, 79]. Based on a map of the training dataset, test data may be added a posteriori. Considering the embedding space instead of the data space may allows to reduce the complexity of data and to mitigate the curse of dimensionality [155], while also reducing the storage capacity required [79]. With this approach, the initial embedding can be obtained using fully supervised Dimensionality Reduction, since class-information is available for the training set.

7.3.2 Parametric Case: Model-Constrained Mapping

Parametric dimensionality reduction methods embed data by applying parametric functions defined on all the data space. Hence, they naturally provide a way of mapping out of sample points. As a result, initially using a model-constrained mapping allows easier incorporation of a posteriori positioning later in the process.

7.3.2.1 Spectral Projection Methods

All algebraic methods (see Sect. 5.1) are intrinsically based on a linear projection, either directly defined in a Euclidean space $\mathcal{D}$ or in a theoretical basis expansion space $\mathcal{K}$ associated with a kernel function. Though **PCA** explicitly uses a projection operator applied to the data points coordinates, methods derived from **cMDS** or Kernel **PCA** only rely on the eigendecomposition of a Gram matrix. To avoid recomputing this decomposition for the Gram matrix extended with the out of sample points, the projection (or eigenfunction of the kernel operator) may be approximated by the Nyström formula [16]:

$$\widehat{\Phi}(\xi) = \sum_{j} \langle \xi, \xi_j \rangle_{\mathcal{K}} x_j \mathring{\Lambda}^{-1},$$

with $\langle \bullet, \bullet \rangle_{\mathcal{K}}$ the kernel function, $\mathring{\Lambda}$ the diagonal matrix containing the d highest eigenvalues of the Gram matrix, and the x_i the embedded sample points, obtained with the eigenvectors (see Annex D.7). Hence, this explicit projection only requires the projection of the sample points and the kernel function computing the inner product in the basis expansion space. The approximation relies on the fact that the eigenvectors and eigenvalues of a big enough sub-matrix of the Gram matrix are close from those of the whole matrix (for the dimensions they have in common), namely that the addition of the out of sample points would not impact significantly the eigendecomposition.

7.3.3 Non-parametric Stress with Neural Network Model

Auto-encoders (see Sect. 5.5.1) are another family of parametric dimensionality reduction techniques. As such it is easy to extend the embedding to other points by applying the encoder part of the artificial neural network to the out of sample data points. This interesting property of auto-encoders has been integrated to the non-parametric method tSNE in [181]. To that end, this method trains an encoder network replacing the loss function constituted of the decoder and reconstruction error, by the stress function of tSNE. As in the case of auto-encoders, the network is pre-trained using restricted Boltzmann machines. Thus, instead of directly performing an optimization on the coordinates of embedded points considered as free parameters, this approach optimizes the stress of tSNE with respect to the weights of the encoder network, the stress being measured for the embedded points obtained by applying the network to the data points.

This approach may be extended to all the non-parametric dimensionality reduction methods that are based on the optimization of a stress function. This extension benefits from the expressive power of artificial neural networks, which enables adaptation to many different types of embeddings. Formally, this corresponds to replacing the optimization problem of Eq. (7.1) by that of Eq. (7.2):

$$X^* = \underset{X \in \mathcal{E}}{\operatorname{argmin}} \, \zeta(X). \tag{7.1}$$

$$X^* = \Phi_{W^*}(\Xi) \text{ with } W^* = \underset{W \in \mathcal{W}}{\operatorname{argmin}} \, \zeta(\Phi_W(\Xi)). \tag{7.2}$$

considering Φ_W the encoder network with weights W.

7.3.4 Non-parametric Case: Mapping-Agnostic Extension

For non-parametric DR techniques, which have not been rendered parametric beforehand, the mapping Φ is only defined for the initial sample $\{\xi_i\}$. Thus, an extension to the entire data space $\mathcal{D}$ may be obtained by interpolation. This leads to the definition of $\widehat{\Phi} : \mathcal{D} \longrightarrow \mathcal{E}$, so that $\Phi(\xi_i) = x_i$ for all points of the initial sample. The embedding space position x associated with the out of sample point ξ is then given by $x = \widehat{\Phi}(\xi)$.

7.3.4.1 Local Linear Transformations: LAMP

Local Affine Multidimensional Projection (LAMP) [93] defines for any out of sample point ξ a local affine transformation $\widehat{\Phi}_\xi$. This transformation is obtained by minimizing for the initial sample points the error between their positions $\Phi(\xi_i)$ obtained with the initial mapping and $\widehat{\Phi}_\xi(\xi_i) = \xi_i A_\xi + b_\xi$ given by this orthogonal affine transformation (where A_ξ is a $\delta \times d$ matrix and b_ξ a $1 \times d$ row vector). This transformation is then applied to the out of sample point providing $x = \widehat{\Phi}_\xi(\xi)$. The locality of this process comes from a weighting of errors depending on the position of ξ. Indeed the weights, associated with the error for the initial point i is computed as:

$$w_i = \frac{\Delta(\xi, \xi_i)^{-2}}{\sum_k \Delta(\xi, \xi_k)^{-2}}.$$

The parameters A_ξ and b_ξ are then obtained by minimizing the error:

$$\zeta^{\mathrm{LAMP}} \triangleq \sum_i w_i \|\xi_i A_\xi + b_\xi - x_i\|_2^2,$$

subject to the orthogonality constraint $A_\xi^\top A_\xi = I_{d \times d}$. This is solved as an orthogonal Procrustes problem relying on singular value decomposition. Hence, each out of sample point ξ may be positioned a posteriori with a complexity of $O(n)$, with n the initial sample size.

7.3.4.2 Manifold Reconstruction

Manifold reconstruction [111] may be seen as extending the discrete mapping as an homeomorphism between the data manifold $\mathcal{M}$ and the embedding space $\mathcal{E}$. This relies on the assumption of manifold learning considering that data lives in a manifold whose intrinsic dimensionality correspond to the embedding dimensionality d. In that case, the restriction of $\widehat{\Phi}$ to the manifold $\mathcal{M}$ may constitute a bijection between $\mathcal{M}$ and $\mathcal{E}$. In order to approximate the manifold, this technique first constructs a simplicial mesh across the embedding space (namely a triangular or tetrahedral mesh for a two- or three-dimensional embedding space). The vertices of that mesh are denoted $\mathring{x}_i$. Then, the inverse $\Psi \triangleq \Phi^{-1}$ of the initial mapping Φ is interpolated on the vertices of this mesh using the Sibson interpolation (also known as natural neighbour interpolation). This interpolation

$\widehat{\Psi}$ generates a mesh with vertices $\{\mathring{\xi}_i\} = \widehat{\Psi}(\mathring{x}_i)$ that approximates the manifold in the data space. Once this reconstruction of the manifold obtained, each out of sample point is orthogonally projected on its "nearest cell" of the high dimensional mesh. Considering that the manifold was well-sampled by the initial points, and that out of sample points are lying scattered around the manifold by a relatively small noise, this cell should be reasonably close. The projected points are then mapped to points of the embedding space using barycentric interpolation. Namely, the high dimensional point is expressed as a barycentre of the cells vertices, and its embedded representative is computed as the barycentre with same weights of associated vertices in the embedding space.

7.3.4.3   Radial Basis Functions Interpolation

Radial Basis Functions (RBF) [49, 148] provide an efficient tool for interpolating a function from a multidimensional space or from any metric space. Hence, they are very well-suited for extending the initial mapping to the entire data space, as proposed by Amorim et al. [5] for the specific case of mapping interaction. In addition, RBF interpolation is robust to data space of high dimensionality. Indeed, the space is sampled as a grid with n_0 steps in each direction, the interpolant converges towards the function with spectral accuracy, meaning that the error is bounded by a term proportional to $(1/n_0)^\delta$ when $n_0 \rightarrow \infty$ [13]. With RBF interpolation, the extension $\widehat{\Phi}$ of the mapping may be expressed by:

$$\widehat{\Phi}(\xi) = \sum_{k=1}^{n} w_k \gamma \left(\frac{\Delta(\xi, \xi_k)}{\sigma_k} \right) + \sum_{k=n+1}^{n+m} w_k p_k(\xi).$$

The weights $\{w_k\}$ with $w_k \in \mathbb{R}^d$ are data-dependent. The weights $\{w_k\}$ for $k \in [\![1, n]\!]$ are associated to each of the n points $\{\xi_k\}$ of the initial sample. The basis functions are positive definite univariate functions $\gamma : \mathbb{R}^+ \longrightarrow \mathbb{R}^+$. Often used

Table 7.1 Basis functions for RBF interpolation

Function name	Function expression $\gamma(u)$
Gaussian	$\exp\left(\frac{-u^2}{2} \right)$
Multiquadric	$\sqrt{1 + u^2}$
Inverse multiquadric	$\frac{1}{\sqrt{1+u^2}}$
Linear	u
Cubic	u^3
Thin plate spline	$u^2 \log u$
C^0 Matérn	$\exp(-u)$
C^1 Matérn	$\exp(-u)(1 + u)$
C^2 Matérn	$\exp(-u)(3 + 3u + u^2)$

basis functions are listed in Table 7.1. In addition, the RBF interpolation may be associated with a low-degree polynomial. Weights $\{w_k\}$ for $k \in [\![n+1, n+m]\!]$ are associated to the m polynomials $\{p_k\}$ with $p_k \in \mathbb{R}[\xi]$ a polynomial with variable ξ. These constitute a basis of the polynomials of a given degree on $\mathbb{R}^\delta$. Note that those polynomials are defined when the high dimensional space is $\mathbb{R}^\delta$, whereas for abstract metric spaces only a polynomial of degree zero may be considered.

The interpolation function $\widehat{\Phi}$ is non-linear with respect to ξ. Yet, the weights $\{w_k\}$ may be determined by solving the following linear system, combining interpolation conditions and constraints on the polynomial terms:

$$
\begin{cases}
\widehat{\Phi}(\xi_1) = x_1, \\
\dots \\
\widehat{\Phi}(\xi_i) = x_i, \\
\dots \\
\widehat{\Phi}(\xi_n) = x_n,
\end{cases}
\quad \text{and} \quad
\begin{cases}
\sum_{k=1}^{n} w_k\, p_1(\xi_k) = 0, \\
\dots \\
\sum_{k=1}^{n} w_k\, p_i(\xi_k) = 0, \\
\dots \\
\sum_{k=1}^{n} w_k\, p_m(\xi_k) = 0,
\end{cases}
\tag{7.3}
$$

where $\{\xi_i\}$ are the n initial sample points and $\{x_i\}$ their embedding space positions.

This linear system may be rewritten in matrix form as:

$$
\left[\begin{array}{c|c} \Gamma_\Xi & P \\ \hline P^\top & 0 \end{array} \right] W = \left[\begin{array}{c} X \\ 0 \end{array} \right].
$$

In this Equation, Γ_Ξ is the square matrix whose element (i, j) is $\gamma\left(\Delta_{ij}/\sigma_j\right)$, P_Ξ is the $n \times m$ matrix whose element (i, j) is $p_j(\xi_i)$, W is the $n+m \times d$ unknown matrix of weights whose i^{th} row is the weight w_i and X is the matrix of embedded points. Using a uniform value σ for the scale parameter, the matrix Γ_Ξ is symmetric, which allows more efficient resolution of the linear problem.

Then, the position x of any out of sample point ξ can be obtained as $x = \widehat{\Phi}(\xi)$. That computation requires a complexity of $O(n)$. For a set of out of sample points defined by a matrix Ξ' of size $(N - n) \times \delta$, the embedding may be obtained with a matrix operation:

$$
X' = \left[\Gamma_{\Xi'\Xi} \mid P_{\Xi'} \right] W,
$$

where $\Gamma_{\Xi'\Xi}$ is the $(N - n) \times n$ matrix whose element (i, j) is $\gamma(\Delta(\xi_i', \xi_j)/\sigma_j)$ and $P_{\Xi'}$ is the $(N - n) \times m$ matrix whose element (i, j) is $p_j(\xi_i')$.

In Fig. 7.5, we present the RBF extension of a mapping obtained with **Class-NeRV** for 512 samples of the digits dataset. Those samples are chosen using Hastie sampling described in [85, 114], which corresponds to a greedy resolution of the k-medoids problem. A Gaussian kernel is used with a uniform scale parameter σ set as the average distance to the 5th neighbour. One of the advantages of Gaussian basis functions is that they may be shown to provide super-spectral convergence in terms of accuracy, as opposed to spectral convergence for other basis functions

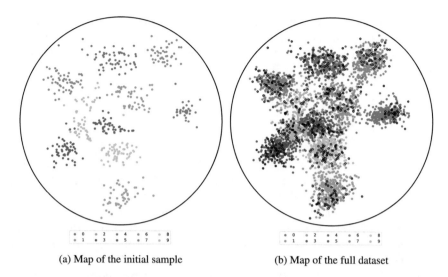

(a) Map of the initial sample   (b) Map of the full dataset

Fig. 7.5 Out of sample extension with RBF interpolation of a mapping of the digits dataset obtained with ClassNeRV. Panel (**a**) shows the map of the initial sample of 512 points, while panel (**b**) presents the map of the entire dataset (3 823 data points) obtained by positioning a posteriori the out of sample points. Black markers on panel (**b**) identify the positions of the initial sample

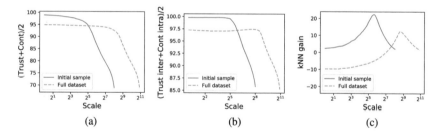

(a)   (b)   (c)

Fig. 7.6 Quality indicators for the mapping of the initial sample, and of the entire dataset with a posteriori positioning. The indicators may not be compared scale by scale between the two curves due to the difference of normalization. (**a**) Structure preservation. (**b**) Class preservation. (**c**) Class purity

such as Multiquadric and Inverse Multiquadric [68]. A polynomial part of degree 0 is used, leading to a matrix $P = \mathbf{1}_{1 \times n}$. With this constant term, the interpolation is robust to rotations, symmetries, scaling and translations of the embedded points. Note that without any polynomial part, RBF interpolation is unable to reproduce a constant function [49].

Based on these maps, the extension of the mapping seems to mostly preserve the classes structure obtained with the initial sample. However, some overlaps appear between the classes. This leads to a loss of map quality as shown Fig. 7.6. This phenomenon may come from the fact that in the initial mapping process no space has been allocated for the additional points in the inter-cluster gap. Thus, some fine-tuning of the global structure obtained by a posteriori positioning might be required to maximize the final map quality.

7.3.5 Intermediate Conclusions

In this work, we mostly focused on mapping quality, relying on multi-scale optimization. Yet, acceleration methods are only defined for methods focusing on the preservation of small scale information. On the other hand, out of sample extension allows to add more points, but may require some fine-tuning of the map if the initial distribution of points is not sufficiently representative. Hence, future work may provide an accelerated version of the multiscale approach relying on out of sample extensions. Indeed, at the highest scale, a small sub-sample of landmark points could be embedded. Then, as the scale progressively decreases (dividing the considered neighbourhood size), points could be progressively added by out of sample extension (multiplying the number of points). Thus, the effective number in the considered neighbourhood of each point i would remain small, allowing to compute only a small number of pairwise forces.

Chapter 8
Applications of Dimensionality Reduction to the Diagnosis of Energy Systems

The results of my long observations and reflections are recorded in the little volume which I now offer to Your Highness: and although I deem this work unworthy of Your Highness's notice, yet my confidence in your humanity assures me that you will accept it, knowing that it is not in my power to offer you a greater gift than that of enabling you to understand in the shortest possible time all those things which I have learnt through danger and suffering in the course of many years. I have not sought to adorn my work with long phrases or high-sounding words or any of those allurements and ornaments with which many writers seek to embellish their books, as I desire no honour for my work but such as its truth and the gravity of its subject may justly deserve.

The prince, Niccolò Machiavelli

This Chapter presents a few examples of applications of dimensionality reduction for the analysis of data towards the diagnosis of energy systems. These systems encompass smart-buildings (Sect. 8.1), photovoltaic systems (Sect. 8.2) and batteries (Sect. 8.3). Diagnosis aims at identifying the occurrence of faults in a system. These faults are characterized by their *signatures,* that is their effects on the system and the monitored variables. The discriminability between the signatures of different faults is a necessary condition for the possibility of diagnostic [64]. In that regard, Dimensionality Reduction (DR) may allow to compare different signatures, provided for instance by I–V curves for photovoltaic systems and by Power Spectral Density of acoustic signals for Li-ion batteries.

8.1 Smart Buildings Commissioning

Buildings are complex systems constituted of many components allowing to maintain indoor comfort conditions such as correct temperature and ventilation. Malfunctions of those components or of their control may lead to significant energy losses or to violations of the desired comfort conditions. With the development of smart buildings, more and more sensors are incorporated to monitor the behaviour of different sub-systems. The data generated by this monitoring provide an opportunity to detect and diagnose faults, thus allowing to perform components maintenance or control correction. Yet, these data are not always exploited at their full potential. This Section describes a technique to present potential fault behaviours in a visual and intelligible way using Dimensionality Reduction. We originally introduced this approach in Geoffroy et al. [77].

8.1.1 System and Rules

The present case study considers the monitoring of a simple air-handling unit. The diagram of that system is shown in Fig. 8.1. This unit handles the ventilation of a room, supplying air from the outdoor environment using a supply fan and extracting indoor air with an exhaust fan. In order to reduce energy loss, a cross-flow heat exchanger mitigates the temperature difference between these two air flows. In winter, the supply air temperature may be increased using a heating coil (powered by a heat pump). In summer, the room temperature may be lowered by free-cooling, relying solely on the fans.

That air handling unit is monitored with five temperature sensors, which measure the temperatures T_{out} (outside air), T_{prh} (pre-heated air), T_{sup} (supply air), T_{in} (return air) and T_{exh} (exhaust air). Additional monitoring indicates whether power is supplied or not to the active elements (fans and heat pump) through Boolean variables P_{fan} and P_{hp}. For this system, we have data for nearly a year with a time step $\Delta t = 1h$, leading to 8512 measures points. These points are separated between a winter period and a summer period (each containing 4346 samples). Each of these period imply a different behaviour for the thermal regulation (i.e. mostly heating in winter and cooling in summer). To estimate the proper behaviour of the system and of its control, several expert rules have been established. Each rule attempts to detect a specific type of fault, by combining the values of different sensor measures.

The conditions indicating a fault are listed in Tables 8.1 and 8.2. Inequality conditions of the form $u(t) > u_0$ (for each subrule i, k) and associated with a scaling factor $u_{90\%}$ are converted into a measure of fault severity at each time t:

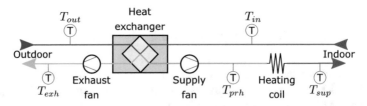

Fig. 8.1 Diagram of the monitored air-handling unit managing the ventilation and thermal regulation of a smart-building

Table 8.1 Rules detecting faults in the air-handling unit behaviour (control or sub-system failure) during the winter period. Temperature thresholds are set to $T_{low} = 19\,°C$ and $T_{high} = 21\,°C$

Fault	Fault condition	$u_{90\%}$
Fault 1	$T_{exh}(t) > T_{sup}(t)$	$2\,°C$
	$T_{in}(t) > T_{out}(t)$	$5\,°C$
Fault 2	$T_{sup}(t) > T_{in}(t) + 5\,°C$	$10\,°C$
Fault 3	$T_{out}(t) > T_{sup}(t)$	$5\,°C$
Fault 4	$T_{low} > T_{in}(t)$	$1\,°C$
	$T_{prh}(t) > T_{sup}(t)$	$1\,°C$
Fault 5	$T_{in}(t) > T_{high}$	$2\,°C$
	$T_{sup}(t) > T_{prh}(t)$	$1\,°C$
Fault 6	$T_{in}(t) > T_{sup}(t) + 1\,°C$	$1\,°C$
	$P_{hc}(t) = 0$	NA

Table 8.2 Rules detecting faults in the air-handling unit behaviour during the summer period. The temperature threshold is set to $T_{high} = 22\,°C$. $\bar{T}_k(t)$ and $\Delta T_k(t)$ are the average and difference between two successive samples (t and $t + \Delta t$) for the kth measure of temperature

Fault	Fault condition	$u_{90\%}$
Fault 1	$T_{sup}(t) > T_{prh}(t)$	$1\,°C$
Fault 2	$T_{in}(t) > T_{out}(t)$	$1\,°C$
	$T_{in}(t) > T_{high}$	$1\,°C$
	$P_{fan}(t) = 0$	NA
Fault 3	$|\Delta T_k(t)| < 10^{-6}$	$\bar{T}_k(t)$

$$h(t) = \tanh\left(\frac{u(t) - u_0}{u_{90\%}} \tanh^{-1}(0.9)\right).$$

A rule h_j defined by several sub-rules h_{jk} is given by a fuzzy equivalent of the logical and operation:

$$h_j(t) = \min_k h_{jk}(t).$$

The fault severity is a value in $]-1, 1[$. Values lower than zero are considered as the absence of fault, so that the effective severity may be assessed by the positive part of that indicator.

For the winter period, there are six types of faults (Table 8.1): Fault 1 corresponds to a malfunction of the heat exchanger. Fault 2 means that the supplied air is excessively warm. Fault 3 indicates heating components failure or wrong sensors implantation. Fault 4 means that the heat pump is not activated despite the room temperature being lower than the minimum comfort value. Fault 5 indicates that the heat pumps warms up the supply air, while the room temperature is above the maximum comfort value. Fault 6 corresponds to the case where the heat pump is not working while air significantly colder than the room temperature is supplied by the air-handling unit. For the summer period, there are three types of faults (Table 8.2). Fault 1 indicates that the heat pump is in the wrong mode (i.e. heating in the summer). Fault 2 means that the free-cooling functionality is not working properly. Fault 3 corresponds to sensors saturation. Those ad hoc rules allow to provide a proof of concept for our commissioning system. Yet, in a real case scenario, more complex rules with parameters fine-tuned depending on the effective properties of the building should be established.

8.1.2   Mapping

At each time step, every severity index is assessed, thus defining a point in a six-dimensional space (in winter) and in a three-dimensional space (in summer). For both cases, a representative subset of 1024 data points is obtained with Hastie sampling procedure [85, 114], which corresponds to a greedy resolution of the k-medoids clustering problem. This sub-sampling allows more tractable computations and a better readability of the map. The points are then mapped in a two-dimensional space using ASKI (see Sect. 5.3.7) with trade-off $\tau = 0.5$ (with mixture type 1), perplexity $\kappa^\star = 64$ and map scale parameter $s_i = 1$ (i.e. non-metric behaviour). The resulting maps for the winter and summer rules are respectively shown Figs. 8.2 and 8.3.

On these maps, the distribution of faults are presented by the pie chart markers, with black markers indicating the absence of faults. For a given pie, the proportion of the ith pie wedge illustrates the proportion η_i of fault i severity in the cumulated severity of all faults. On the other hand, this cumulated fault severity h is translated by the pie area (the area being a more reliable graphic variable than the radius for that encoding [153]). The proportions $\{\eta_i\}$ and cumulated severity h are formally expressed as:

$$\eta_i(t) \triangleq \frac{h_i(t)^+}{\sum_k h_k(t)^+} \quad \text{and} \quad h(t) \triangleq \sum_k h_k(t)^+.$$

Considering the map of winter rules (Fig. 8.2) we see that the points for which no fault has been detected are scattered near the center. Markers radii tend to increase when moving away from that central area, so that points subject to more severe cumulated faults are presented closer to the map border. As for the fault types, they appear to be organized around that center, with regions associated to a given main fault. Counter-clockwise, we find regions of faults 2, 4, 6, 1 and 5 (fault 3 never being the main fault). Interfaces between two regions, show a progressive transition between one fault and another. Hence, these interfaces allow to observe co-occurrences of different faults. Hence, the radial position in the map indicates the cumulated severity of faults, whereas the angular position translates the fault types.

As for the map of summer rules (Fig. 8.3), it shows a clear separation between two clusters. Indeed, in practice, the third rule behaves as a Boolean variable, thus inducing the strong gap between the two clusters (with and without Fault 3). In the

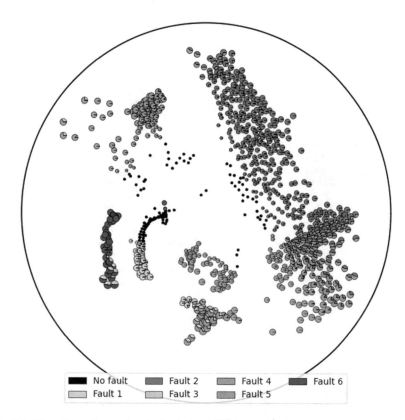

Fig. 8.2 Map of the winter rules obtained with ASKI for 1024 sub-sampled points. Markers are pie charts showing the proportion of each fault (positive part of fault indicators), while the area shows the severity (sum of those positive parts)

right cluster (without Fault 3), the points with no fault regroup in the lower right part with Fault 1 at the top, Fault 2 on the left, and a slight interfaces between the two. Similarly to winter rules, the cumulated severity increases when moving away from the no fault region. In the left cluster (with Fault 3), co-occurrences of Faults 1 and 2 may be observed, which are closer from the main regions of those faults. Finally, more results may be found in [76] concerning the application of Dimensionality Reduction for the monitoring of smart-buildings.

8.2 I–V Characterization of Photovoltaic Systems

Photovoltaics systems (PV) convert solar irradiance into direct current electrical power, often converted to alternating current through an inverter. A common non-destructive method for the on-site diagnosis of these systems relies on the measure

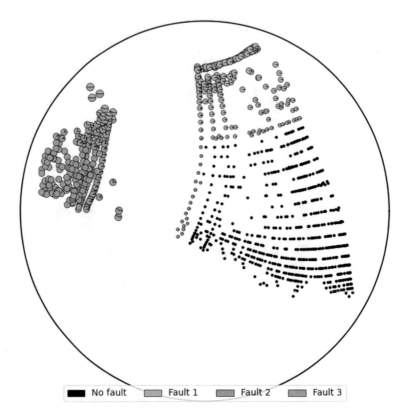

Fig. 8.3 Map of the summer rules obtained with ASKI for 1024 sub-sampled points. Markers are defined similarly to Fig. 8.2

of an I–V curve. It allows, for instance, to identify several types of faults impacting the parameters of the electrical model.

8.2.1 I–V Curves

A photovoltaic (PV) cell, module or array of modules may be characterized by its I–V curve, namely the relation between its output voltage V and current I. The I–V curve may be measured either in a laboratory, with standardized "flash tests", or on site during operation of the PV system. The shape of an I–V curve depends on the characteristics of the module, but also on environmental conditions such as the irradiance G or temperature T. A common approach to estimate the I–V curve of a PV system (module or array) combines in series and/or parallel the individual I–V characteristics of many PV cells, along with a few other components such as bypass diodes. A PV cell may be represented, with the one-diode model [191],

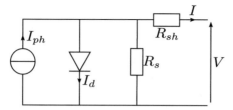

Fig. 8.4 One diode model of a photovoltaic cell (electrical diagram)

as the combination of a current source, a diode and series and shunt resistances. The electrical diagram of a cell in that model is represented Fig. 8.4. It leads to the following relation between the current I and V for a cell:

$$I = I_{ph} - I_0 \left(\exp \left(-\frac{e(V + R_s I)}{n k_B T} \right) - 1 \right) - \frac{V + R_s I}{R_p}, \tag{8.1}$$

where I_{ph} is the photo-generated current, I_0 and n the dark saturation and ideality factor characterizing the diode, R_s and R_p the series and shunt resistances. The photo-generated current I_{ph} varies with the irradiance G and temperature T. The constants e and k_B are the electrical charge of an electron and the Boltzmann's constant.

In the common topology, PV modules are constituted of cells connected in series. Hence, for a given current, the voltage of the module is obtained by adding the voltages of the cells for that current. Ideally, all cells of a module should be identical and receive the same irradiance. In that case, the I–V curve of the module may be obtained from the I–V curve of the cells by multiplying the voltage by the number of cells, leading to the type of curve observed in Fig. 8.5a. The intercepts of this curve define two characteristic variables: the open-circuit voltage V_{oc} (voltage for $I = 0$) and the short-circuit current I_{sc} (current for $V = 0$). The curve may also be used to identify the maximum power point of the PV system, with voltage V_{mpp} and current I_{mpp} maximizing the output power $P = VI$. During the operation of the system, voltage is regulated to reach this point using maximum power point tracking algorithms.

In practice, the I–V characteristic may vary between cells, causing a mismatch. This variability may come from the cells properties (i.e. differences in manufacturing or ageing), or from the environmental conditions (e.g., partial shading of the modules). In the presence of mismatch, the output current of the module is constrained by the characteristics of the worst cells. Indeed, for a connection in series the current is the same in all cells. These cells may be reversed biased to allow higher values of current. This means that they have a negative voltage, inducing power consumption and heating. To avoid this situation, bypass diodes are connected in parallel in PV systems, each diode allowing to short-circuit a specific group of cells. Thus, in the presence of mismatch, I–V curves appear as shown

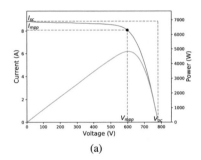
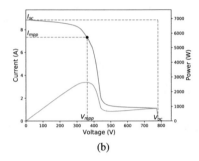

(a)   (b)

Fig. 8.5 Example I–V curve (blue) and associated power output (orange) for a module without and with mismatch. Dashed lines show the open-circuit voltage V_{oc}, short-circuit current I_{sc}, and maximum power point voltage and current V_{mpp} and I_{mpp}. (**a**) Without mismatch. (**b**) With mismatch

Fig. 8.5b. Mismatch significantly impact the performances of the module with a lower value of the maximum output power.

8.2.2 Comparing Normalized I–V Curves

We consider the case of I–V curves measured onto several power plants or obtained by simulation. To alleviate the effect of environmental conditions, such as irradiance and temperature, and of architecture differences, the voltage and current values are respectively normalized by V_{oc} and I_{sc}. Those normalized curves are embedded in a two-dimensional space using **ClassNeRV** (see Sect. 6.6) favouring the tears ($\tau^{\in} = 0.5$ and $\tau^{\notin} = 0$) and with a non-metric behaviour ($s_i = 1$). The labels provided to the algorithm indicate the power plant for which the curve has been measured. The resulting map is shown Fig. 8.6, with point markers representing the corresponding I–V curves. In this map, there are only a few overlaps between the INES power plant, Sacarine power plant and simulated curves. It indicates that the three groups have different characteristics, which could be explained by the difference of modules technology and/or ageing. In addition, in the INES power plant, the map coloring distinguishes two different arrays INES-1 and INES-2, both regrouped in a common INES class in the labels, so that this distinction is not provided to the algorithm. The map also informs on the different behaviours of those two arrays.

Finally, focusing on the bottom of the map and considering the shape of the curves, we can see that the curves with mismatch are isolated from the rest. Moreover, different types of mismatches can be identified. Indeed, if the part with high voltage and low current is long, as in the bottom right curves of INES-2, it means that there is a great proportion of the PV array with lower performances. On the contrary, the bottom left simulated curves correspond to arrays for which only a small proportion of the array has lower performances. Note that the current level of that part of the curve indicates at the severity of the mismatch, namely the difference

Fig. 8.6 Map of the
normalized curves with
non-metric ClassNeRV
($\tau^{\in} = 0.5$ and $\tau^{\neq} = 0$)

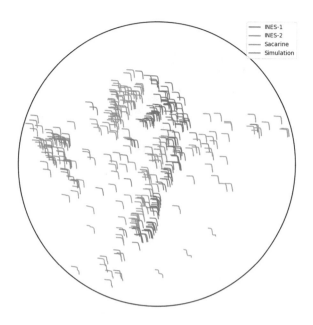

of performance between the normal cells and the limiting cells. This severity also seem to impact the repartition of curves in the map.

To assess the reliability of the information given by the map, we display MING graphs (see Sect. 4.3) in Fig. 8.7. Those graphs tend to show that the map is mostly reliable at the considered scale. Yet, a strong tear appears between the curves from the Sacarine power plant on the right of the map and the INES and simulated curves. This may indicate that the structure is too complicated to be well-represented in two dimensions. Another significant distortion appears for the bottom right curves with mismatch (from INES), which are rightfully placed close from one another, but should be closer from the simulated curves with mismatch at the bottom left.

Since I–V curves of different arrays are mostly associated to different areas of the map, we can conclude that these arrays have distinct behaviours. This could mean that the technology of modules constituting each array are different, or that these arrays have been subject to different ageing. In the case of the INES power plant, the second cause is more probable the two arrays have been installed at the same time with similar modules. In addition, the simulated I–V curves are represented apart in the map. Hence, it shows that they are not very representative of the real cases in this dataset. As such, the choice of parameters for these simulations could be challenged to represent these cases.

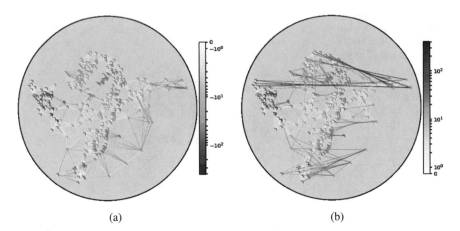

Fig. 8.7 MING graphs for the ClassNeRV map of normalized I–V curves with $\kappa = 10$ and trustworthiness/continuity penalizations. (**a**) Retrieval graph. (**b**) Relevance graph

8.2.3   Colour Description of the Chemical Compositions

A PV cell is made of silicon. However, the chemical composition of the cell may locally vary because of pollution and other impurities of the material. These pollutions should be avoided because they may decrease the yield of the cell. Spectrometry is sometimes used to study the local chemical composition. However, for a given area of the cell, the spectrometry provides many measures for the various wavelengths. The spectra associate to each analyzed area can then be seen as a high dimensional data. Of course, it is uneasy to have a global analysis along the cell from such a dataset.

Dimensionalityt reduction can help to infer from such data. In view of this goal, we test here a metric mapping method (see Sect. 5.2.1) and a non-metric method (Sect. 5.2.2). Proximity in the resulting maps shows the proximity in terms of chemical composition. The maps are associated to the HSV colour space (c.f. Sect. 1.3.5.1) [113]. This allows associating a colour to each local measure: the colour proximity now codes for a proximity between measures. Each measure describes the local chemical properties of a given area of the cell and is associated to a colour. Consequently, a colour picture results from such an analysis: different colours highlight the different types of pollution. Areas with close colours corresponds to areas where the chemical properties are close.

We can observe on the map reached by the metric method (here, Isomap—see (Sect. 5.1.3.2) that most of the analysed surface is shown as close to cyan colours (Fig. 8.8). Several small areas are colored in red, dark blue or white. Conversely, a much higher variety of colours may be found in the non-metric method's map

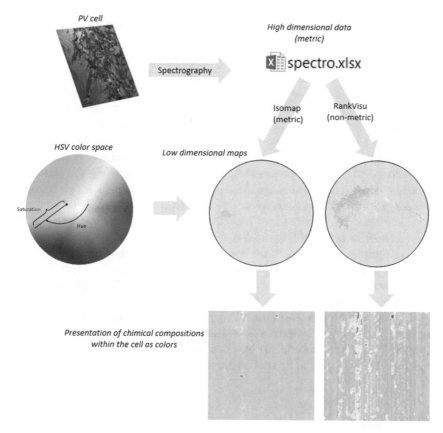

Fig. 8.8 MSpectrometries along a PV cell gives access to the local chemical composition. The measurement on each cell's area may been seen as a point in a high dimensional space. The set of measurements on every analysed area of the cell is presented here as an Excel file. This dataset in embedded in a 2D space thanks to two mapping methods. one metric and one non-metric. Maps may be associated with the 2D HSV colour space as done in ColorPhylo method [113]. This process allows associating each analyzed area of the cell with a unique colour that codes for proximity in terms of chemical composition

(here, RankVisu—see Sect. 5.2.2.1). This is due to the different behaviors of these methods as detailed in the list of points below:

- The metric tmethods account for distances. Here, several clearly different compositions have been observed. These highly different areas stretch the colourcode and tend to hide the small variability of composition.
- Conversely, the non-metric method accounts for neighborhood relations. This tends to reduce the contribution of outliers, which allows presenting a finer description of small variations.

In the following Sections we address a few other applications.

8.3 Acoustic Characterization of Li-ion Batteries

Electrochemical storage systems are key components for the development of intermittent energy sources and electrical vehicles. A possible non-destructive technique for monitoring the internal state of a battery during operation relies on the measure of acoustic signals at its surface as detailed in [83]. Such monitoring may be categorized into Ultrasound Characterization (UsC) and Acoustic Emission (AE) as illustrated Fig. 8.9. Ultrasound Characterization is an active technique requiring the emission of an acoustic signal on the surface of the battery. Depending on the internal mechanical properties of the battery, the transmission, absorption and reflection of the signal may change locally. Therefore, the output signal is impacted by those properties. Conversely, Acoustic Emission is a passive approach where the signal is emitted by the release of local mechanical constraints inside the battery (e.g., gas bubbles in electrolyte, cracks).

To analyse the signals recorded during monitoring, many descriptors may be computed, such as the signal amplitude, energy or peak frequency. Here, we focus on the Power Spectral Density (PSD), which characterizes the frequency content of a signal for a range of frequencies up to 1000 kHz. The PSD may be computed based on the Fast Fourier Transform using the method described in [198]. PSD are sequence data that may be treated as multidimensional data, considering the value of the PSD for each frequency as the coordinate along an axis of the data space. Thus, the dimensionality of this data space is 1024.

8.3.1 Case 1

We first consider Ultrasound Characterization signals obtained during Vincent Gau PhD training for a Li-ion battery. The PSD of those signals are embedded in two dimensions using ASKI with trade-off $\tau = 0.5$ and the non-metric scaling ($s_i = 1$), leading to the map shown Fig. 8.10. The State of Charge (SoC) of the battery

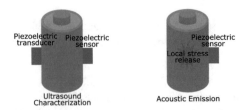

Fig. 8.9 Measurement set-up for Ultrasound Characterization and Acoustic Emission. An acoustic signal is measured at the surface of the battery using a piezoelectric sensor. This signal may be emitted at the surface of the battery with a piezoeletric transducer (in the case of UsC) or spontaneously engendered due to inner mechanical constraints (in the case of AE). This Figure is inspired from [83]

Fig. 8.10 ASKI map of the
UsC, with state of charge
added in colour

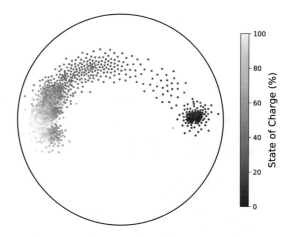

associated to each measure of acoustic signals is added in colour on the point
makers. This map tends to show that there exists a relation between the frequency
content of the measured acoustic signals (represented by the position in the map)
and the SoC of the battery. This could probably be explained by the variation of the
electrodes density due to lithium insertion. Thus, this map seems to indicate that for
this type of battery, Ultrasound Characterization could be used to predict the State
of Charge. Furthermore, the map also shows a distinct behaviour for very low SoC.

8.3.2 Case 2

In this second case, we focus on the PSD of Acoustic Emission signals, measured
by Florence Degret during her PhD training. Those signals have been measured for
a Li-ion battery. These signals are embedded in two dimensions using ASKI with
trade-off $\tau = 0.5$ and the non-metric scaling ($s_i = 1$). The resulting map is shown
Fig. 8.11, with the maximum value of the PSD represented in colour. This shows
that signals with higher amplitudes are mapped to the left of the map. In this map, a
few outliers occupy most of the space. For the sake of map readability, those outliers
are discarded, thus considering only the points in the dashed rectangle.

To study the reliability of the map, MING graphs (see Sect. 4.3) are displayed
Fig. 8.12. On those graphs, the bottom left area appears reliably represented. Yet, the
other areas are subject to significant clutter. For the retrieval graph shown Fig. 8.12a,
the zooms in the denser areas show a mix of white and blue edges, indicating that
the structure is probably glued. For the relevance graph, we successively apply
interactive edge filtering to study two map clusters, corresponding to modes of the
points distribution. The first focus shows that the top cluster is mostly linked with
itself and a little linked to the top of the bottom cluster. Similarly, the second focus

Fig. 8.11 Map of the Power Spectral Density obtained with ASKI. Maximum value of the PSD is added in colour (with a logarithmic colour scale). The dashed rectangle shows the subset of points considered to increase map readability by discarding the five outliers

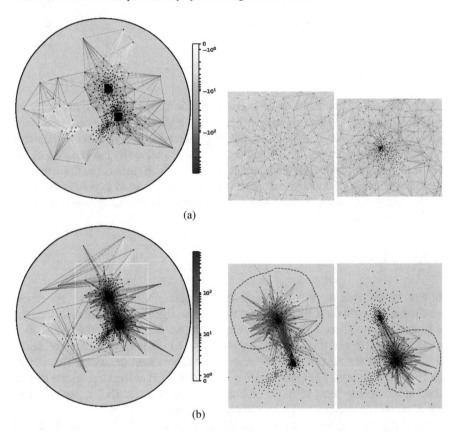

Fig. 8.12 MING graphs for the map of the PSD obtained with ASKI (without outliers), with $\kappa = 10$, and using trustworthiness and continuity indicators. (**a**) Retrieval graph. (**b**) Relevance graph both complete and with interactive edge filtering successively focusing on two map clusters

indicate that the bottom cluster is mostly linked to itself and a little linked with the top of the top cluster. In addition those two clusters are not linked with the bottom left area. As a consequence, these clusters are not ideally represented locally, but their is no misleading tear at the global level.

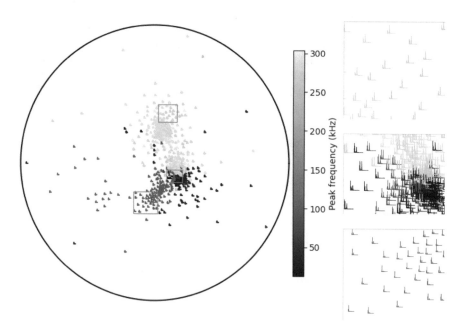

Fig. 8.13 Map of the Power Spectral Density obtained with ASKI, with five outliers removed (as shown Fig. 8.11). Peak frequency is presented in colour. Points are represented by their associated PSD

Figure 8.13 presents the positioning of different spectra, with the peak frequency shown in colour. The peak frequency indicates for which frequency the PSD is maximum. Note that the point markers are replaced here by the PSD normalized by its maximum value (previously shown Fig. 8.11), so that different signals may be represented identically. The map organization appear strongly related to the peak frequency, which indicates that the signals tend to be separated based on the position of theirs peaks. In the top area, the PSD have a main peak at a rather high frequency, whereas in the bottom area, they have a different main peak at a lower frequency. A transition area is also identified where two peaks of close amplitude coexist. In that area, the peak frequency strongly vary since small variations may reverse the order between the two peaks and thus lead to a very different peak frequency. This shows that the representation with a map gives a finer categorization of the signals than the summary by a simple ad hoc scalar descriptor.

Dimensionality Reduction allows to represent complex measures characterizing the state of a system through a map. As such, we showed its interest for several applications. Indeed, it allowed to present expert indicators for the supervision of a smart-building thermal regulation, to identify differences of behaviours between several PV arrays, to show a link between the SoC of a battery and its acoustic characterization, and to present the discriminability between different acoustic measures based on their frequency content. These preliminary applications could be extended in many ways. For instance, other metrics could be considered to compare specific signals (e.g., I–V curves or acoustic signals), such as the L_1 distance, Dynamic Time Warping or geodesic distances in those ambient metric spaces. Moreover, such signals could be confronted with effective annotations of faults. In that case, ClassNeRV could be applied to help identify the discriminability between these faults based on the considered features.

Chapter 9
Conclusions

The thorough study of nature is the most ground for mathematical discoveries.

Joseph Fourier

What we know is not much. What we do not know is immense.

Pierre-Simone de Laplace

Dimensionalityt Reduction (DR) enables analysts to perform visual exploration of high dimensional data by providing a low-dimensional representation. On its own, it allows, for instance, to identify at a glance data structures such as clusters, hierarchies or outliers . Through the use of colours or marker shapes, it can also help to forge hypotheses about relations between the data features and additional variables, either categorical (classes) or quantitative. This book reviewed the state-of-the-art for the analysis of multidimensional data and proposed new tools for a DR based study of a dataset, as explained below.

In order to obtain an embedding of a dataset, ASKI (Sect. 5.3.7) offers an adaptive behaviour depending on the discrepancy between the intrinsic dimensionality and embedding dimensionality. It focuses on the preservation of neighbourhoods, but also of distances if possible, that is if the dimensionality discrepancy is low. Thus, it limits the occurrence of excessive gaps between clusters that may be observed with tSNE for data of low or intermediate dimensionality.

To assess the effective complexity of the dataset, TIDLE (Sect. 2.3) proposes a local evaluation of the intrinsic dimensionality of data based on the distribution of distances to the two nearest neighbours of each point. It relies on the assumption that the intrinsic dimensionality is locally uniform for a considered neighbourhood size.

The embedding process may also benefit from available class-information to improve the representation of classes. As such, class-guided neighbourhood embedding techniques, i.e. ClassNeRV and ClassJSE (Sect. 6.6), account for class labels to steer unavoidable distortions so as to reduce their impact on the representation of classes. To this end, they penalize more false neighbours within classes and missed neighbours between classes.

© The Author(s), under exclusive license to Springer Nature Switzerland AG 2022
S. Lespinats et al., *Nonlinear Dimensionality Reduction Techniques*,
https://doi.org/10.1007/978-3-030-81026-9_9

A similar paradigm is implemented by class-aware indicators (Sect. 3.3.3) for map evaluation. Indeed, they assess the preservation of classes by measuring specifically intraclass false neighbours and interclass missed neighbours. Thereby, they relate the preservation of classes with the effective data structure instead of focusing solely on class purity in the map.

In order to inform the analyst about potential representation errors, MING (Sect. 4.3) allows to locate the distortions in a map based on the display of nearest neighbours graphs. These graphs present reliable neighbourhood relations as well as false and missed neighbours. The severity of these distortions is assessed based on pre-existing rank-based map quality indicators. As such, they explain locally the overall value of those indicators.

To position a posteriori points on an existing map, RBF interpolation (Sect. 7.3.4.3) allows to extend any mapping, either parametric or non-parametric. The quality of this interpolation is robust to increase in the data dimensionality.

This book also briefly presented some applications of multidimensional data analysis for the diagnosis of energy systems. First, DR was used to show the co-occurrence of faults detected by a set of expert rules for smart buildings (Sect. 8.1). Then, I–V curves measured for PV power plants were mapped (Sect. 8.2), allowing to detect differences of behaviour between different plants, as well as to isolate curves subject to mismatch problems. Finally, the Power Spectral Density of acoustic signals obtained for the diagnosis of batteries were considered (Sect. 8.3). In the context of ultrasound characterization, it showed a link between the frequency content of signals and the state of charge of the battery in the context of ultrasound characterization. For acoustic emission signals, the two-dimensional representation of those frequency contents based on a map showed a finer characterization of the variability of signals than ad hoc signal descriptors, such as peak frequency.

Appendix A
Some Technical Results

> *No human investigation can be called real science it is cannot be demonstrated mathematically.*
>
> Leonardo Da Vinci
>
> *My first mistress is the Mathematics, and she is a jealous, jealous mistress.*
>
> Ben H. Rogers

A.1 Equivalence Between Triangle Inequality and Convexity of Balls for a Pseudo-Norm

The equivalence between the triangle inequality of the pseudo-norm and the convexity of balls for this pseudo-norm is stated Sect. 1.1.4. It may be proven as follows.

Assume triangle inequality of the pseudo norm $||.||$. Let $\xi_1, \xi_2 \in \mathcal{B}$ the ball centered at the origin of radius ω for $||.||$ and $\tau \in [0, 1]$. The triangle inequality gives:

$$||\tau\xi_1 + (1 - \tau)\xi_2|| \leqslant ||\tau x|| + ||(1 - \tau)\xi_2||$$

The homogeneity of the pseudo-norm leads to:

$$||\tau\xi_1 + (1 - \tau)\xi_2|| \leqslant |\tau|||\xi_1|| + |1 - \tau|||\xi_2||$$

ξ_1 and ξ_2 being in the unit ball, their norm is lower than ω:

$$||\tau\xi_1 + (1 - \tau)\xi_2|| \leqslant \tau\omega + (1 - \tau)\omega = \omega$$

Thus, $\tau\xi_1 + (1 - \tau)\xi_2$ is in $\mathcal{B}$, which means that $\mathcal{B}$ is a convex set.

Assume convexity of any ball $\mathcal{B}(\omega)$ for the pseudo-norm $||.||$. Let $\xi_1, \xi_2 \in \mathcal{D} \setminus \{0\}$. $\frac{\xi_1}{||\xi_1||}, \frac{\xi_2}{||\xi_2||} \in \mathcal{B}(1)$ (based on the homogeneity property) and $\frac{||\xi_1||}{||\xi_1||+||\xi_2||} + \frac{||\xi_2||}{||\xi_1||+||\xi_2||} = 1$, thus using the definition of convexity, the weighted mean of

© The Author(s), under exclusive license to Springer Nature Switzerland AG 2022
S. Lespinats et al., *Nonlinear Dimensionality Reduction Techniques*,
https://doi.org/10.1007/978-3-030-81026-9

elements of the unit ball belongs again to the unit ball, namely

$$
\left\| \frac{\|\xi_1\|}{\|\xi_1\| + \|\xi_2\|} \frac{\xi_1}{\|\xi_1\|} + \frac{\|\xi_2\|}{\|\xi_1\| + \|\xi_2\|} \frac{\xi_2}{\|\xi_2\|} \right\| \leqslant 1
$$

Which leads using homogeneity to:

$$
\|\xi_1 + \xi_2\| \leqslant \|\xi_1\| + \|\xi_2\|.
$$

An example showing that the $L_{1/2}$ pseudo-norm (see Sect. 1.1.4) does not satisfy the triangle inequality, can be obtained by taking $\xi_1 = (1, 0)$ and $\xi_2 = (0, 1)$. This leads to:

$$
\|\xi_1 + \xi_2\|_{1/2} = (\sqrt{1} + \sqrt{1})^2 = 4 > \|\xi_1\|_{1/2} + \|\xi_2\|_{1/2} = (\sqrt{1} + \sqrt{0})^2 + (\sqrt{0} + \sqrt{1})^2 = 2.
$$

A.2   From Pareto to Exponential Distribution

The logarithm of a variable following a Pareto distribution of parameter ∂ follows an exponential distribution of parameter ∂. This result used in Sect. 2.3 may be proven as follows.

Let X and Y be random variables, so that X follows a Pareto distribution with Probability Density Function (PDF) $p_X(x) = \partial x^{-\partial - 1}$, and $Y = \log X$. Considering the transformation g and its inverse g^{-1} defined by $g(x) = \exp(x)$ and $g^{-1}(y) = \log(y)$, the PDF of Y is given by:

$$
p_Y(y) = p_X\left(g^{-1}(y)\right) \left| \frac{dg^{-1}(y)}{dy} \right|
$$

$$
= \partial \exp(y)^{-\partial - 1} \exp(y)
$$

Thus $p_Y(y) = \partial \exp(-\partial y)$, which is the PDF of an exponential distribution with parameter ∂.

A.3   Spiral and Swiss roll

The Archimedian spiral used in Sect. 1.2.3.1 is parametrized by the following equations with free parameter t:

$$
\begin{cases}
\xi_1 = t \cos(t), \\
\xi_2 = t \sin(t).
\end{cases}
$$

It may be extruded along the ξ_3 axis to form a **Swiss roll** with free parameters t and ξ_3.

Yet, a uniform distribution of the variable t does not provide uniform distribution along the spiral. If we denote $f(t) = (\xi_1(t), \xi_2(t))$, the curvilinear abscissa x is given by:

$$x(t) = \int_0^t \left\| \frac{df(u)}{du} \right\| du.$$

For the spiral, $f(t) = (t\cos(t), t\sin(t))$, thus:

$$\left\| \frac{df(u)}{du} \right\|^2 = \|(\cos(u) - u\sin(u), \sin(u) + u\cos(u))\|^2$$

$$= (\cos(u) - u\sin(u))^2 + (\sin(u) + u\cos(u))^2$$

$$= \left(\cos^2(u) - \overline{2u\cos(u)\sin(u)} + u^2\sin^2(u) \right)$$

$$+ \left(\sin^2(u) + \overline{2u\cos(u)\sin(u)} + u^2\cos^2(u) \right)$$

$$= 1 + u^2,$$

leading to:

$$\left\| \frac{df(u)}{du} \right\| = \sqrt{1 + u^2},$$

Computing the integral of this quantity along the path leads to the following expression of the curvilinear abscissa:

$$x(t) = \frac{1}{2} \left(\operatorname{arcsinh}(t) + t\sqrt{1 + t^2} \right).$$

Hence, a uniformly sampled spiral may be obtained by taking $x^{-1}(u)$ with u drawn from a uniform distribution.

Appendix B
Kullback–Leibler Divergence

True Laws of Nature cannot be linear.

Albert Einstein

The Kullback–Leibler divergence is at the core of most Neighbourhood Embedding methods (see Sect. 5.3).

B.1   Generalized Kullback–Leibler Divergence

The generalized Kullback–Leibler divergence is used by Class-Guided Neighbourhood Embedding methods to ensure the positivity of their stress despite the fact that the divergence inputs may not be treated as probability distributions (see Sect. 6.6).

A divergence $\mathscr{D}(u, v)$ is positive for all scalar or vector pairs (u, v) and vanishes only for identical arguments $u = v$. For a strictly convex generator function F, the Bregman divergence $\mathscr{D}_F(u, v)$ [24] is defined by the difference between the function evaluated in u and the tangent (hyper-plane) to the function at point v also evaluated in u, giving:

$$\mathscr{D}_F(u, v) = F(u) - F(v) - \langle \nabla F(v), u - v \rangle.$$

Considering that a convex function is always above its tangent, the Bregman divergence is positive and null only for $u = v$. The generalized Kullback–Leibler divergence correspond to Bregman divergence whose generator function is the Shannon entropy taken with minus sign:

$$-H(u) = \sum_i u_i \log(u_i).$$

Indeed:

$$\mathscr{D}_{gKL}(u, v) = \mathscr{D}_{-H}(u, v)$$

$$= \sum_i u_i \log u_i - \sum_i v_i \log v_i - \left\langle (..., \log v_i + 1, ...), (..., u_I - -V_i, ...) \right\rangle$$

$$= \sum_i u_i \log u_i - \sum_i v_i \log v_i - \sum_i (u_i \log v_i - v_i \log v_i + u_i - v_i).$$

Which leads to:

$$\boxed{\mathscr{D}_{gKL}(u, v) = \sum_i \left(u_i \log \left(\frac{u_i}{v_i} \right) + v_i - u_i \right).}$$

We may note that, if u and v are probability distributions, i.e. $\sum_i u_i = \sum_i v_i = 1$, then the generalized Kullback–Leibler divergence gives back the classical expression of the Kullback–Leibler divergence (which may be negative for non-probability distributions):

$$\mathscr{D}_{KL}(u, v) = \sum_i u_i \log \left(\frac{u_i}{v_i} \right).$$

B.1.1   Perplexity with Hard Neighbourhoods

We show here the link between the perplexity measure of the distribution of membership degrees and a number of neighbours (see Sect. 5.3.2), considering hard neighbourhoods.

With hard neighbourhoods, the membership degrees may be expressed using flat kernels by:

$$\beta_{ij} = \frac{1_{\mathbb{R}+}(1 - \Delta_{ij}/\sigma_i)}{\sum_{k \neq i} 1_{\mathbb{R}+}(1 - \Delta_{ik}/\sigma_i)} \quad \text{and} \quad b_{ij} = \frac{1_{\mathbb{R}+}(1 - D_{ij}/s_i)}{\sum_{k \neq i} 1_{\mathbb{R}+}(1 - D_{ik}/s_i)}$$

with $1_{\mathbb{R}+}$ the Heaviside function. The entropy of distribution β_i is given by:

$$H_i = - \sum_{j \neq i} \beta_{ij} \log \beta_{ij}$$

$$= \frac{\sum_{j \neq i | \Delta_{ij} \leq \sigma_i} 1 \cdot \log(\sum_{k \neq i | \Delta_{ik} \leq \sigma_i} 1)}{\sum_{k \neq i | \Delta_{ik} \leq \sigma_i} 1}$$

$$= \log \left| \{j \neq i \mid \Delta_{ij} \leq \sigma_i\} \right|,$$

with $|\cdot|$ the cardinal of a set. The associated perplexity is thus:

$$\kappa_i = \left|\{j \neq i \mid \Delta_{ij} \leqslant \sigma_i\}\right|.$$

This means that, for hard neighbourhoods, the perplexity for a given radius corresponds to the number of neighbours within this radius.

B.2 Link Between Soft and Hard Recall and Precision

The link between soft and hard recall and precision is mentioned in Sect. 5.3.3, and extends to the supervised break-down of these terms (see Sects. 3.3.3 and 6.6).

For assessing the preservation of the neighbourhood of a given point i, we may use:

- Soft recall: $\tilde{\mathcal{R}}_i = \sum_{j \neq i} \beta_{ij} \log\left(\frac{\beta_{ij}}{b_{ij}}\right)$.
- Soft precision: $\tilde{\mathcal{P}}_i = \sum_{j \neq i} b_{ij} \log\left(\frac{b_{ij}}{\beta_{ij}}\right)$.
- Intraclass soft recall: $\tilde{\mathcal{R}}_i^{\in} = \sum_{j \in S_i^{\in}} \beta_{ij} \log\left(\frac{\beta_{ij}}{b_{ij}}\right) + b_{ij} - \beta_{ij}$.
- Interclass soft precision: $\tilde{\mathcal{P}}_i^{\notin} = \sum_{j \in S_i^{\notin}} b_{ij} \log\left(\frac{b_{ij}}{\beta_{ij}}\right) + \beta_{ij} - b_{ij}$.

We may show that those terms are related to the hard indices used for map evaluation: recall, precision, intraclass recall and interclass precision, building on the demonstration in [189]. Let consider nearly hard neighbourhoods in both spaces. The membership degrees for these neighbourhoods are assumed equal to:

$$\beta_{ij} = \begin{cases} \frac{1-\epsilon}{\kappa_i} & \text{if } j \in v_i \\ \frac{\epsilon}{N-1-\kappa_i} & \text{if } j \in v_i^{C} \end{cases} \quad \text{and} \quad b_{ij} = \begin{cases} \frac{1-\epsilon}{k_i} & \text{if } j \in n_i \\ \frac{\epsilon}{N-1-k_i} & \text{if } j \in n_i^{C} \end{cases}$$

where v_i and n_i are the neighbourhoods of point i in the data and embedding spaces, of respective size κ_i and k_i. When $\epsilon \to 0$ the neighbourhoods become hard neighbourhoods. Soft recall may be decomposed as follows:

$$\tilde{\mathcal{R}}_i = \sum_{j \in v_i \cap n_i} \overbrace{\frac{1-\epsilon}{\kappa_i} \log\left(\frac{1-\epsilon}{\kappa_i}\frac{k_i}{1-\epsilon}\right)}^{o\left(\frac{1}{\kappa_i}\log\left(\frac{1-\epsilon}{\epsilon}\right)\right)}$$

$$+ \sum_{j \in v_i^{C} \cap n_i} \overbrace{\frac{\epsilon}{N-1-\kappa_i} \log\left(\frac{\epsilon}{N-1-\kappa_i}\frac{k_i}{1-\epsilon}\right)}^{\xrightarrow[\epsilon \to 0]{} 0}$$

$$+ \sum_{j \in v_i \cap n_i^{\complement}} \underbrace{\frac{1-\epsilon}{\kappa_i} \log \left(\frac{1-\epsilon}{\kappa_i} \frac{N-1-k_i}{\epsilon} \right)}_{\underset{\epsilon \to 0}{\sim} \frac{1}{\kappa_i} \log\left(\frac{1-\epsilon}{\epsilon}\right)}$$

$$+ \sum_{j \in v_i^{\complement} \cap n_i^{\complement}} \underbrace{\frac{\epsilon}{N-1-\kappa_i} \log \left(\frac{\epsilon}{N-1-\kappa_i} \frac{N-1-k_i}{\epsilon} \right)}_{\underset{\epsilon \to 0}{\longrightarrow} 0}$$

Considering only the predominant bottom left term, and knowing that $v_i \cap n_i^{\complement} = v_i \setminus n_i$, this leads to:

$$\tilde{\mathscr{R}}_i \underset{\epsilon \to 0}{\sim} \mathscr{R}_i \log \left(\frac{1-\epsilon}{\epsilon} \right) \quad \text{with} \quad \mathscr{R}_i = \frac{|v_i \setminus n_i|}{\kappa_i}.$$

In the same way, soft precision may be written as:

$$\tilde{\mathscr{P}}_i = \sum_{j \in n_i \cap v_i} \overbrace{\frac{1-\epsilon}{k_i} \log \left(\frac{1-\epsilon}{k_i} \frac{\kappa_i}{1-\epsilon} \right)}^{o\left(\frac{1}{k_i} \log\left(\frac{1-\epsilon}{\epsilon}\right)\right)}$$

$$+ \sum_{j \in n_i^{\complement} \cap v_i} \overbrace{\frac{\epsilon}{N-1-k_i} \log \left(\frac{\epsilon}{N-1-k_i} \frac{\kappa_i}{1-\epsilon} \right)}^{\underset{\epsilon \to 0}{\longrightarrow} 0}$$

$$+ \sum_{j \in n_i \cap v_i^{\complement}} \underbrace{\frac{1-\epsilon}{k_i} \log \left(\frac{1-\epsilon}{k_i} \frac{N-1-\kappa_i}{\epsilon} \right)}_{\underset{\epsilon \to 0}{\sim} \frac{1}{k_i} \log\left(\frac{1-\epsilon}{\epsilon}\right)}$$

$$+ \sum_{j \in n_i^{\complement} \cap v_i^{\complement}} \underbrace{\frac{\epsilon}{N-1-k_i} \log \left(\frac{\epsilon}{N-1-k_i} \frac{N-1-\kappa_i}{\epsilon} \right)}_{\underset{\epsilon \to 0}{\longrightarrow} 0}$$

Thus leading to:

$$\tilde{\mathscr{P}}_i \underset{\epsilon \to 0}{\sim} \mathscr{P}_i \log \left(\frac{1-\epsilon}{\epsilon} \right) \quad \text{with} \quad \mathscr{P}_i = \frac{|n_i \setminus v_i|}{\kappa_i}.$$

These derivations may be simply adapted to intraclass and interclass soft recall and precision, by intersecting the sets of indices with S_i^{ϵ} or $S_i^{\neq}$. The only difference in that case is the Bregman correction (difference between KL and Bregman). Yet, this term is dominated by the logarithmic term, namely:

$$b_{ij} - \beta_{ij} = o\left(\log\left(\frac{1-\epsilon}{\epsilon}\right)\right).$$

Hence, intraclass soft recall and interclass soft precision are equivalent to:

$$\tilde{\mathscr{R}}_i^\in \underset{\epsilon \to 0}{\sim} \sum_{j \in (v_i \cap n_i^C) \cap S_i^\in} \frac{1}{\kappa_i} \log\left(\frac{1-\epsilon}{\epsilon}\right)$$

and

$$\tilde{\mathscr{P}}_i^\notin \underset{\epsilon \to 0}{\sim} \sum_{j \in n_i \cap v_i^C \cap S_i^\notin} \frac{1}{k_i} \log\left(\frac{1-\epsilon}{\epsilon}\right).$$

Thus:

$$\tilde{\mathscr{R}}_i^\notin \underset{\epsilon \to 0}{\sim} \mathscr{R}_i^\notin \log\left(\frac{1-\epsilon}{\epsilon}\right) \quad \text{and} \quad \tilde{\mathscr{P}}_i^\notin \underset{\epsilon \to 0}{\sim} \mathscr{P}_i^\notin \log\left(\frac{1-\epsilon}{\epsilon}\right).$$

Equivalent results may be obtained for $\tilde{\mathscr{R}}_i^\notin$ and $\tilde{\mathscr{P}}_i^\in$.

In practice, neighbourhood membership degrees are not of this form, with different values accounting for different levels of severity of the false and missed neighbours (all accounted equivalently by recall and precision). Hence, the real soft recall and precision may be linked with trustworthiness and continuity.

Appendix C
Details of Calculations for Stress Function and Gradients

We present here the derivation of some equations used in the algorithms presented in the manuscript.

C.1 General Gradient of Stress Function

For a given stress function ζ depending solely on embedding distances, the gradient according to the coordinates of one embedded point x_k is:

$$\frac{\partial \zeta}{\partial x_k} = \sum_i \sum_{j \neq i} \frac{\partial \zeta}{\partial D_{ij}} \frac{\partial D_{ij}}{\partial x_k}$$

Since a distance D_{ij} only depends on the positions of x_i and x_j, if $i \neq k$ and $j \neq k$

$$\frac{\partial D_{ij}}{\partial x_k} = 0,$$

leading to

$$\frac{\partial \zeta}{\partial x_k} = \sum_{j \neq k} \underbrace{\frac{\partial \zeta}{\partial D_{kj}} \frac{\partial D_{kj}}{\partial x_k}}_{F_{kj}^*} + \sum_{i \neq k} \underbrace{\frac{\partial \zeta}{\partial D_{ik}} \frac{\partial D_{ik}}{\partial x_k}}_{F_{ik}^*}.$$

For an embedding space with Euclidean distance, and by defining the forces $F_{ij}^* = \frac{\partial \zeta}{\partial D_{ij}}$ it gives:

$$\frac{\partial \zeta}{\partial x_k} = \sum_{j \neq k} F_{kj}^* \frac{x_k - x_j}{D_{kj}} - \sum_{i \neq k} F_{ik}^* \frac{x_i - x_k}{D_{ik}}, \tag{C.1}$$

so that the gradient vector may be computed by adding the sum on rows and on columns of the following order three tensor of size $N \times N \times d$ whose entry (i, j, k) is:

$$F_{ij}^* \frac{X_{ik} - X_{jk}}{D_{ij}}.$$

For every method using a stress function of the form $\zeta = \sum_k \zeta_k$, forces F_{ij}^* could be computed as:

$$F_{ij}^* = \frac{\partial \zeta}{\partial D_{ij}} = \sum_k \frac{\partial \zeta_k}{\partial D_{ij}}. \tag{C.2}$$

Considering the symmetry of distances, the opposite of the gradient (used in gradient descent optimization) may be written as:

$$-\frac{\partial \zeta}{\partial x_k} = \sum_{j \neq k} (F_{kj}^* + F_{jk}^*) \frac{x_j - x_k}{D_{kj}},$$

with $F_{kj} = F_{kj}^* + F_{jk}^*$ the algebraic value of the force exerted upon x_k by a point x_j and $\frac{x_j - x_k}{D_{kj}}$ the unit vector determining the direction of this force. For positive values of F_{kj} the force is attractive, whereas for negative values, it is repulsive.

We may note that if the forces are symmetric, as for most Multi-Dimensional Scaling stresses (e.g., CCA [54] and SNLM [158]), or for symmetrized neighbourhood embedding stresses (e.g., tSNE stress [183]), the forces simplify to $F_{kj}^* + F_{jk}^* = 2F_{kj}^*$, so that:

$$\frac{\partial \zeta^{\text{sym}}}{\partial x_k} = 2 \sum_{j \neq k} F_{kj}^* \frac{x_k - x_j}{D_{kj}}.$$

C.2   Neighbourhood Embedding

In this Section, we compute the forces F_{ij}^* for neighbourhood embedding methods.

C.2.1 Supervised Neighbourhood Embedding (Asymmetric Case)

For classical NeRV [189] and JSE [109], the point-wise stress associated with point k is:

$$\zeta_k = \mathcal{M}_{\mathrm{KL},12}(b_k, \beta_k, \tau).$$

The supervised version (which leads back to the classical version, for single-valued trade-off parameter $\tau_{ij} = \tau$) is similar, using τ_k as a distribution and the generalized Kullback–Leibler formulation of the mixture:

$$\zeta_k = \mathcal{M}_{\mathrm{gKL},12}(b_k, \beta_k, \tau_k),$$

with $\tau_i = (\tau_{ij})_{j \in [\![1;N]\!]}$.

Using Eq. (C.2) for NeRV and JSE, we obtain for the force F_{ij}^*:

$$
\begin{aligned}
F_{ij}^* &= \sum_k \frac{\partial \mathcal{M}_{\mathrm{gKL},12}(b_k, \beta_k, \tau_k)}{\partial D_{ij}} \\
&= \sum_k \sum_{m \neq k} \frac{\partial \mathcal{M}_{\mathrm{gKL},12}(b_k, \beta_k, \tau_k)}{\partial b_{km}} \underbrace{\frac{\partial b_{km}}{\partial D_{ij}}}_{0 \text{ if } k \neq i} \\
&= \sum_{m \neq i} \frac{\partial \mathcal{M}_{\mathrm{gKL},12}(b_i, \beta_i, \tau_k)}{\partial b_{im}} \frac{\partial b_{im}}{\partial D_{ij}},
\end{aligned}
\tag{C.3}
$$

where $\frac{\partial b_{km}}{\partial D_{ij}} = 0$ for $k \neq i$ only applies in the non-symmetrized version of neighbourhood embedding methods.

Using the soft-min expression of output neighbourhood membership degrees b_{ij} of Eq. (C.10), the differential with respect to distances is:

$$
\frac{\partial b_{im}}{\partial D_{ij}} = \begin{cases} a_{ij}' b_{ij}(b_{im} - 1), & \text{if } m = j, \\ a_{ij}' b_{ij} b_{im}, & \text{if } m \neq j. \end{cases}
\tag{C.4}
$$

Hence, using Eqs. (C.3) and (C.4), we get:

$$
F_{ij}^* = -a_{ij}' b_{ij} \sum_{m \neq i} b_{im} \left(\frac{\partial \mathcal{M}_{\mathrm{gKL},12}(b_i, \beta_i, \tau_i)}{\partial b_{ij}} - \frac{\partial \mathcal{M}_{\mathrm{gKL},12}(b_i, \beta_i, \tau_i)}{\partial b_{im}} \right),
\tag{C.5}
$$

which leads to (for efficient computation):

$$F_{ij}^* = a_{ij}' \left[b_{ij} \sum_{m \neq i} \left(b_{im} \frac{\partial \mathcal{M}_{\mathrm{gKL},12}(b_i, \beta_i, \tau)}{\partial b_{im}} \right) - \left(b_{ij} \frac{\partial \mathcal{M}_{\mathrm{gKL},12}(b_i, \beta_i, \tau_i)}{\partial b_{ij}} \right) \right]$$

$$(C.6)$$

The computation of b_{ij} is detailed Sect. C.2.3, that of a_{ij}' is presented Sect. C.2.4, and Sect. C.2.2 shows, for the different types of mixtures, the computation of:

$$b_{ij} \frac{\partial \mathcal{M}_{gKL}(b_i, \beta_i, \tau_i)}{\partial b_{ij}}.$$

C.2.2   Mixtures

In the following, we denote the scalar Bregman divergence (which is the elementary term in the computation of the generalized Kullback–Leibler divergence):

$$\mathcal{D}_{\mathrm{B}}(\beta_{ij}, b_{ij}) = \beta_{ij} \log\left(\frac{\beta_{ij}}{b_{ij}} \right) + b_{ij} - \beta_{ij}$$

C.2.2.1   Type 1

For (Class)NeRV, the mixture of type 1 of the generalized Kullback–Leibler divergence is:

$$\mathcal{M}_{\mathrm{gKL},1}(b_i, \beta_i, \tau_i) = \sum_{j \neq i} \tau_{ij} \mathcal{D}_{\mathrm{B}}(\beta_{ij}, b_{ij}) + (1 - \tau_{ij}) \mathcal{D}_{\mathrm{B}}(b_{ij}, \beta_{ij})$$

Leading to the following differentiation for $j \neq i$:

$$b_{ij} \frac{\partial \mathcal{M}_{\mathrm{gKL},1}(b_i, \beta_i, \tau_i)}{\partial b_{ij}} = b_{ij} - \beta_{ij} + (1 - \tau_{ij}) \mathcal{D}_{\mathrm{B}}(b_{ij}, \beta_{ij}) \qquad (C.7)$$

C.2.2.2 Type 2

For (Class)JSE, the mixture of type 2 of the generalized Kullback–Leibler divergence is:

$$\mathcal{M}_{\text{gKL},2}(b_i, \beta_i, \tau_i) = \sum_{j \neq i} \frac{1}{\tau_{ij}} \mathcal{D}_{\text{B}}(\beta_{ij}, \mu_{ij}) + \frac{1}{1 - \tau_{ij}} \mathcal{D}_{\text{B}}(b_{ij}, \mu_{ij})$$

with $\mu_{ij} = \tau_{ij} b_{ij} + (1 - \tau_{ij}) \beta_{ij}$.

We may note that for the type 2 mixture, there is no need for the Bregman adaptation since:

$$\mathcal{M}_{\text{gKL},2}(b_i, \beta_i, \tau_i) = \sum_{j \neq i} \frac{1}{\tau_{ij}} \left(\beta_{ij} \log \left(\frac{\beta_{ij}}{\mu_{ij}} \right) + \overbrace{\mu_{ij} - \beta_{ij}}^{\tau_{ij}(b_{ij} - \beta_{ij})} \right)$$

$$+ \sum_{j \neq i} \frac{1}{1 - \tau_{ij}} \left(b_{ij} \log \left(\frac{b_{ij}}{\mu_{ij}} \right) + \underbrace{\mu_{ij} - b_{ij}}_{(1 - \tau_{ij})(\beta_{ij} - b_{ij})} \right)$$

$$= \sum_{j \neq i} \frac{1}{\tau_{ij}} \beta_{ij} \log \left(\frac{\beta_{ij}}{\mu_{ij}} \right) + \frac{1}{1 - \tau_{ij}} b_{ij} \log \left(\frac{b_{ij}}{\mu_{ij}} \right),$$

namely:

$$\boxed{\mathcal{M}_{\text{gKL},2}(b_i, \beta_i, \tau_i) = \mathcal{M}_{\text{KL},2}(b_i, \beta_i, \tau_i)}.$$

This gives the following differentiation for $j \neq i$:

$$\boxed{b_{ij} \frac{\partial \mathcal{M}_{\text{gKL},2}(b_i, \beta_i, \tau_i)}{\partial b_{ij}} = b_{ij} - \beta_{ij} + \frac{1}{1 - \tau_{ij}} \mathcal{D}_{\text{B}}(b_{ij}, \mu_{ij})} \qquad \text{(C.8)}$$

C.2.2.3 Asymptotic Behaviour

The two terms of the mixture of type 2 are not defined for $\tau_{ij} = 1$ or $\tau_{ij} = 0$ respectively, but they converge to a limit that we may compute using the following expansions:

$$\frac{1}{1 - \tau_{ij}} b_{ij} \log \left(\frac{b_{ij}}{\mu_{ij}} \right) = \frac{b_{ij}}{1 - \tau_{ij}} \log \left(1 + \overbrace{\underbrace{\frac{b_{ij} - \mu_{ij}}{\mu_{ij}}}_{\underset{\tau_{ij} \to 1}{\longrightarrow} b_{ij}}}^{(1 - \tau_{ij})(b_{ij} - \beta_{ij})} \right) \underset{\tau_{ij} \to 1}{\sim} \frac{b_{ij}}{1 - \tau_{ij}} \frac{1 - \tau_{ij}}{b_{ij}} (b_{ij} - \beta_{ij})$$

$$\frac{1}{1-\tau_{ij}}(\mu_{ij}-b_{ij}) = \frac{(\tau_{ij}-1)b_{ij}+(1-\tau_{ij})\beta_{ij}}{1-\tau_{ij}} = \beta_{ij}-b_{ij}$$

$$\boxed{\frac{1}{1-\tau_{ij}}\mathscr{D}_B(b_{ij},\mu_{ij}) \xrightarrow[\tau_{ij}\to 1]{} 0}$$

$$\frac{1}{\tau_{ij}}\beta_{ij}\log\left(\frac{\beta_{ij}}{\mu_{ij}}\right) = \frac{\beta_{ij}}{\tau_{ij}}\log\left(1+\overbrace{\underbrace{\frac{\beta_{ij}-\mu_{ij}}{\mu_{ij}}}_{\xrightarrow[\tau_{ij}\to 0]{}\beta_{ij}}}^{\tau_{ij}(\beta_{ij}-b_{ij})}\right) \underset{\tau_{ij}\to 0}{\sim} \frac{\beta_{ij}}{\tau_{ij}}\frac{\tau_{ij}}{\beta_{ij}}(\beta_{ij}-b_{ij})$$

$$\frac{1}{\tau_{ij}}(\mu_{ij}-\beta_{ij}) = \frac{\tau_{ij}b_{ij}-\tau_{ij}\beta_{ij}}{\tau_{ij}} = b_{ij}-\beta_{ij}$$

$$\boxed{\frac{1}{\tau_{ij}}\mathscr{D}_B(\beta_{ij},\mu_{ij}) \xrightarrow[\tau_{ij}\to 0]{} 0}$$

Thus, if all τ_{ij} tend to one, the two types of mixtures converge to the same limit, which is the soft recall with generalized Kullback–Leibler divergence:

$$\lim_{\tau_{ij}\to 1}\mathscr{M}_{\text{gKL},2}(b_i,\beta_i,\tau_i) = \lim_{\tau_{ij}\to 1}\mathscr{M}_{\text{gKL},1}(b_i,\beta_i,\tau_i)$$

$$= \sum_{j\neq i}\beta_{ij}\log\left(\frac{\beta_{ij}}{b_{ij}}\right)+b_{ij}-\beta_{ij}\,.$$

In the same way, when all τ_{ij} tend to 0, the mixtures both converge to soft precision with generalized Kullback–Leibler divergence:

$$\lim_{\tau_{ij}\to 0}\mathscr{M}_{\text{gKL},2}(b_i,\beta_i,\tau_i) = \lim_{\tau_{ij}\to 0}\mathscr{M}_{\text{gKL},1}(b_i,\beta_i,\tau_i) =$$

$$\sum_{j\neq i}b_{ij}\log\left(\frac{b_{ij}}{\beta_{ij}}\right)+\beta_{ij}-b_{ij}$$

For the derivatives, we have:

$$\lim_{\tau_{ij}\to 1}b_{ij}\frac{\partial\mathscr{M}_{\text{kl},2}(b_i,\beta_i,\tau_i)}{\partial b_{ij}} = \lim_{\tau_{ij}\to 1}b_{ij}\frac{\partial\mathscr{M}_{\text{kl},2}(b_i,\beta_i,\tau_i)}{\partial b_{ij}} = b_{ij}-\beta_{ij}\,.$$

$$\lim_{\tau_{ij}\to 0}b_{ij}\frac{\partial\mathscr{M}_{\text{kl},2}(b_i,\beta_i,\tau_i)}{\partial b_{ij}} = \lim_{\tau_{ij}\to 0}b_{ij}\frac{\partial\mathscr{M}_{\text{kl},2}(b_i,\beta_i,\tau_i)}{\partial b_{ij}} = b_{ij}\log\left(\frac{b_{ij}}{\beta_{ij}}\right)\,.$$

As a result, the type 2 mixture may be extended continuously for $\tau_{ij} = 1$ or $\tau_{ij} = 0$ using the equivalent values for type 1 mixture.

C.2.3   Membership Degrees

Neighbourhood membership degrees are defined for point i as $\beta_i = (\beta_{ij})_{j \neq i}$ (in the data space) and $b_i = (b_{ij})_{j \neq i}$ (in the embedding space). The terms β_{ij} and b_{ij} are computed using soft-min of arguments α_{ij} and a_{ij} (defined in Sect. C.2.4):

$$\beta_{ij} = \frac{\exp(-\alpha_{ij})}{\sum_{k \neq i} \exp(-\alpha_{ik})}, \tag{C.9}$$

$$b_{ij} = \frac{\exp(-a_{ij})}{\sum_{k \neq i} \exp(-a_{ik})}. \tag{C.10}$$

For numerical computations with floating point algebra, high values of the arguments may lead to all exponential terms reaching zero. In this case of numerical underflow, the soft-min is zero divided by zero. To avoid this phenomenon, the soft-min may be stabilized as follows, so that at least one of the exponential term is non-zero:

$$\beta_{ij} = \frac{\exp\left(-\alpha_{ij} + \min_{k \neq i} \alpha_{ik}\right)}{\sum_{k \neq i} \exp\left(-\alpha_{ik} + \min_{k \neq i} \alpha_{ik}\right)},$$

$$b_{ij} = \frac{\exp\left(-a_{ij} + \min_{k \neq i} a_{ik}\right)}{\sum_{k \neq i} \exp\left(-a_{ik} + \min_{k \neq i} a_{ik}\right)}.$$

C.2.4   Soft-Min Arguments

For each point i the input arguments α_{ij} are computed applying the function α_i (negative logarithm of the kernel function) to the distances Δ_{ij} scaled by σ_i:

$$\alpha_{ij} \triangleq \alpha_i \left(\frac{\Delta_{ij}}{\sigma_i}\right)$$

Their derivatives may be computed with respect to σ_i in order to set the scale by perplexity (see Sect. C.2.5):

$$\alpha'_{ij} \triangleq \frac{\partial \alpha_{ij}}{\partial \sigma_i} = -\frac{\Delta_{ij}}{\sigma_i^2} \alpha'_i \left(\frac{\Delta_{ij}}{\sigma_i}\right).$$

Similarly, the output arguments a_{ij} may be obtained by applying a_i (negative logarithm of the kernel function) on distances D_{ij} scaled by s_i:

$$a_{ij} \triangleq a_i \left(\frac{D_{ij}}{s_i} \right)$$

Their derivatives are computed with respect to D_{ij} in order to optimize the stress function of the mapping process:

$$a'_{ij} \triangleq \frac{\partial a_{ij}}{\partial D_{ij}} = \frac{1}{s_i} a'_i \left(\frac{D_{ij}}{s_i} \right)$$

Several types of kernels are available:

C.2.4.1   Gaussian Kernel

For the Gaussian kernel, the argument is:

$$- \log \mathcal{N}(u) = \frac{u^2}{2} \quad \text{and} \quad (- \log \mathcal{N})'(u) = u$$

$$\alpha_{ij} = \frac{\Delta_{ij}^2}{2\sigma_i^2} \quad \text{and} \quad \alpha'_{ij} = -\frac{\Delta_{ij}^2}{\sigma_i^3}$$

$$a_{ij} = \frac{D_{ij}^2}{2s_i^2} \quad \text{and} \quad a'_{ij} = \frac{D_{ij}}{s_i^2} \tag{C.11}$$

C.2.4.2   Student Kernel

Student kernel with degree of freedom λ_i in the data space or l_i in the embedding space:

$$- \log \mathscr{S}_{\lambda_i}(u) = \frac{\lambda_i + 1}{2} \log \left(1 + \frac{u^2}{\lambda_i} \right) \quad \text{and} \quad (- \log \mathscr{S}_{\lambda_i})'(u) = \frac{\lambda_i + 1}{\lambda_i} \frac{u}{1 + \frac{u^2}{\lambda_i}}$$

$$\alpha_{ij} = \frac{\lambda_i + 1}{2} \log \left(1 + \frac{\Delta_{ij}^2}{\lambda_i \sigma_i^2} \right) \quad \text{and} \quad \alpha'_{ij} = -\frac{\lambda_i + 1}{\lambda_i} \frac{\Delta_{ij}^2}{\sigma_i^3 + \frac{\Delta_{ij}^2 \sigma_i}{\lambda_i}}$$

$$a_{ij} = \frac{l_i + 1}{2} \log \left(1 + \frac{D_{ij}^2}{l_i s_i^2} \right) \quad \text{and} \quad a'_{ij} = \frac{l_i + 1}{l_i} \frac{D_{ij}}{s_i^2 + \frac{D_{ij}^2}{l_i}} \tag{C.12}$$

C.2.4.3 Asymptotic Behaviour

Using the Taylor expansion of $u \mapsto \log(1 + u)$ at $u = 0$, we may show that for high degrees of freedom, the Student kernel converges to the corresponding Gaussian kernel:

$$\mathscr{S}_{\lambda_i}(u) \underset{\lambda_i \to +\infty}{\sim} \mathscr{N}(u) \text{ and } \mathscr{S}'_{\lambda_i}(u) \underset{\lambda_i \to +\infty}{\sim} \mathscr{N}'(u)$$

C.2.5 Scale Setting by Perplexity

Setting the scale parameter σ_i so that the perplexity of the membership degrees β_i equals $\kappa^\star$ amounts to solving:

$$H(\beta_i) = \log \kappa^\star,$$

where H_i is the Shannon entropy of β_i defined by:

$$\boxed{H_i = - \sum_{j \neq i} \beta_{ij} \log \beta_{ij}}$$

This may be solved using Newton-Raphson root-finding algorithm, in order to find the zero of $H_i - \log \kappa^\star$. Thus, we need to know the derivative of Shannon entropy:

$$\frac{\partial H_i}{\partial \sigma_i} = - \sum_{j \neq i} (\log \beta_{ij} + 1) \frac{\partial \beta_{ij}}{\partial \sigma_i}$$

Using Eq. (C.9) we get:

$$\frac{\partial \beta_{ij}}{\partial \sigma_i} = - \beta_{ij} (\alpha'_{ij} - \sum_{k \neq i} \alpha'_{ik} \beta_{ik})$$

Thus:

$$\frac{\partial H_i}{\partial \sigma_i} = \sum_{j \neq i} \beta_{ij} (\log \beta_{ij} + 1)(\alpha'_{ij} - \sum_{k \neq i} \alpha'_{ik} \beta_{ik})$$

$$= \underbrace{\sum_{j \neq i} \alpha'_{ij} \beta_{ij} (\log \beta_{ij} + 1)}_{} - \underbrace{\sum_{j \neq i} \beta_{ij} (\log \beta_{ij} + 1)}_{-H_i + 1} \underbrace{\sum_{k \neq i} \alpha'_{ik} \beta_{ik}}_{\sum_{j \neq i} \alpha'_{ij} \beta_{ij}}$$

$$\boxed{\frac{\partial H_i}{\partial \sigma_i} = \sum_{j \neq i} \alpha'_{ij} \beta_{ij} (\log \beta_{ij} + H_i)}$$

C.2.6   Force Interpretation

Using Eqs. (C.5) and (C.7), we obtain the following formulation of forces for neighbourhood embedding methods with type 1 mixture (Eq. (C.8)) leading to a very similar formulation for the type 2 mixture):

$$
F_{ij}^* = a_{ij}' \left[\underbrace{\beta_{ij}}_{\boxed{1}} + \underbrace{b_{ij} \sum_{k \neq i} (1 - \tau_{ik}) \mathscr{D}_B(b_{ik}, \beta_{ik})}_{\boxed{2}} - \underbrace{b_{ij}}_{\boxed{3}} - \underbrace{(1 - \tau_{ij}) \mathscr{D}_B(b_{ij}, \beta_{ij})}_{\boxed{4}} \right]
$$

The terms $\boxed{1}$ and $\boxed{2}$ correspond to attractive forces, while $\boxed{3}$ and $\boxed{4}$ correspond to repulsive forces. The term $\boxed{3}$ repulses all the map neighbours of i, while $\boxed{1}$ attracts all the data space neighbours of i (these two combined form a force $\beta_{ij} - b_{ij}$ that is repulsive if the points are too close, attractive if they are too far and null if the membership degrees are equal). They are the only ones considered by SNE and tSNE ($\tau = 1$). The terms $\boxed{2}$ and $\boxed{4}$ depend on the effective divergence from the embedding to the data space. The term $\boxed{4}$ repulses false neighbours (b_{ij} high despite low β_{ij}) especially affecting neighbours that are not from the class of point i (and affecting less the neutral neighbours). Conversely, $\boxed{2}$ attracts all map neighbours, the magnitude of that attraction increasing with the amount of false neighbours surrounding i especially if those are not of the same class as i. These two antagonist forces seem to attract all close by neighbours towards i while its neighbourhood is not well-represented (high total value of divergence) and consistently pushing away the false interclass neighbours, thus selecting reliable and intraclass neighbours.

The weighting by a_{ij}' implies that a point i affects another point j more when the change of distance Δ_{ij} has a significant impact on the argument a_{ij}. When the embedding space kernel is a Student kernel, a_{ij}' tends to zero when the distance D_{ij} tends towards infinity (see Eq. (C.12)), leading to a plastic behaviour (the forces vanish between far points). Conversely, for a Gaussian kernel, a_{ij}' increases linearly with the distance D_{ij} (see Eq. (C.11)).

Appendix D
Spectral Projections Algebra

We present here the algebraic details of the resolution of some spectral projection methods detailed in Sect. 5.1.

D.1 PCA as Matrix Factorization and SVD Resolution

PCA may be viewed as a matrix factorization method (see Sect. 5.1.1.2). In that framework, it approximates a $N \times \delta$ centered data matrix $\tilde{\Xi}$ by the product of two low rank matrices, namely a matrix X of size $N \times d$ of coordinates in the embedding space and a matrix of archetypes A of size $d \times \delta$ (coordinates of archetypes in the canonical basis of the data space), minimizing the reconstruction error:

$$\zeta^{\text{PCA}} = ||\tilde{\Xi} - XA||_2$$

with $|| \bullet ||_2$ the standard Frobenius norm.

This leads to a SVD based resolution of the minimization problem. For any $N \times \delta$ data matrix $\tilde{\Xi}$ of rank $r \leqslant \min(N, \delta)$, there exist a Singular Value Decomposition (SVD) so that

$$\tilde{\Xi} = U\sqrt{\Lambda}V^{\top} \tag{D.1}$$

with U a $N \times r$ orthogonal matrix (i.e. $U^{\top}U = I_{r \times r}$), V a $\delta \times r$ orthogonal matrix (i.e. $V^{\top}V = I_{r \times r}$) and $\sqrt{\Lambda}$ an $r \times r$ a real positive diagonal matrix containing the singular values of $\tilde{\Xi}$ sorted in descending order of magnitude.

© The Author(s), under exclusive license to Springer Nature Switzerland AG 2022
S. Lespinats et al., *Nonlinear Dimensionality Reduction Techniques*,
https://doi.org/10.1007/978-3-030-81026-9

The orthogonal invariance of the Frobenius norm gives:

$$||\tilde{\Xi} - XA||_2 = ||\sqrt{\Lambda} - \underbrace{U^\top XAV}_{\sqrt{\widehat{\Lambda}}}||_2 \tag{D.2}$$

Due to the rank constraints on X and A the maximum possible rank of $\sqrt{\widehat{\Lambda}}$ is d. To minimize the reconstruction error subject to this constraint, $\sqrt{\widehat{\Lambda}}$ must be a diagonal matrix containing the d first diagonal values of $\sqrt{\Lambda}$ and zeros for the $r - d$ other diagonal elements, namely:

$$\sqrt{\widehat{\Lambda}} = \sqrt{\Lambda} I_{r \times d} I_{d \times r},$$

with $I_{r \times d}$ and $I_{d \times r}$ the pseudo-diagonal identity matrices of size $r \times d$ and $d \times r$. Hence, $I_{r \times d} I_{d \times r}$ is a $r \times r$ block matrix, with a $I_{d \times d}$ upper left block and zeros everywhere else.

Based on the definition of $\sqrt{\widehat{\Lambda}}$ in Eq. (D.2), we obtain:

$$XA = U\sqrt{\widehat{\Lambda}}V^\top \tag{D.3}$$
$$= U\sqrt{\Lambda} I_{r \times d} I_{d \times r} V^\top$$

The product XA may be factored in many different ways. By choosing the solution with orthonormal archetypes, namely by imposing the constraint $AA^\top = I_{d \times d}$, we get:

$$X = U\sqrt{\Lambda} I_{r \times d}, \tag{D.4}$$

$$A = I_{d \times r} V^\top. \tag{D.5}$$

D.2   Link with Linear Projection

The matrix factorization problem may also be interpreted as finding a linear projection on a subspace of low dimensionality d, defined by a matrix $PP^\top$ with P of size $\delta \times d$. This projection minimizes:

$$\zeta^{PCA} = ||\tilde{\Xi} - \tilde{\Xi} PP^\top||_2,$$

which is equivalent to the previous formulation by posing $\tilde{\Xi} PP^\top = XA$. This matrix contains the coordinates of the projected points in the data space.

Using the SVD of $\tilde{\Xi}$ (Eq. (D.1)), in Eq. (D.4) we get:

$$X = \underbrace{U\sqrt{\Lambda}V^\top}_{\tilde{\Xi}} \underbrace{VI_{r\times d}}_{A^\top},$$

So that both matrices X and A may be computed only the matrix V using Eq. (D.5) and:

$$X = \tilde{\Xi}A^\top. \tag{D.6}$$

Using this results, we may also show the equivalence of the matrix formulation and linear projection formulations of **PCA**:

$$XA = \tilde{\Xi} \underbrace{\overbrace{VI_{r\times d}}^{A^\top} \overbrace{I_{d\times r}V^\top}^{A}}_{PP^\top},$$

which leads to:

$$P = A^\top.$$

We may note that P does the projection in the embedding space basis (or basis of the projection subspace), and $PP^\top$ the projection in the canonical basis of the data space.

D.3   Sparse Expression

Left multiplication by $I_{r\times d}$ and right multiplication by $I_{d\times r}$ respectively correspond to selection of the d first rows and columns of a matrix. Thus, only part of the elements of matrices U, V and $\sqrt{\Lambda}$ are required for the computations. Considering that

$$\sqrt{\Lambda}I_{r\times d} = I_{r\times d}I_{d\times r}\sqrt{\Lambda}I_{r\times d},$$

we may rewrite the Eqs. (D.4), (D.6) and (D.5) using smaller matrices $\mathring{U} = UI_{r\times d}$, $\mathring{V} = VI_{r\times d}$ and $\sqrt{\mathring{\Lambda}} = I_{d\times r}\sqrt{\Lambda}I_{r\times d}$. This gives:

$$\boxed{X = \mathring{U}\sqrt{\mathring{\Lambda}} = \tilde{\Xi}\mathring{V}}$$

$$\boxed{A = \mathring{V}^\top}$$

$\mathring{U}$ of size $N \times d$ and $\mathring{V}$ of size $\delta \times d$ contain the d first columns of U and V, and $\sqrt{\mathring{\Lambda}}$ of size $d \times d$ contains the d first rows and columns of $\sqrt{\Lambda}$.

D.4   PCA and Centering: From Affine to Linear

PCA may be seen as simultaneously performing a linear regression estimating $\tilde{\Xi}$ from X (optimizing A), and optimizing X so as to get the minimum modelling error. Some approaches for solving PCA optimize alternately those two parameters.

We may show that the best linear projection or regression on centered data $\tilde{\Xi}$ is equivalent to the best affine projection or regression on the original data Ξ as mentioned in Sect. 5.1.1. For any matrix X, the best affine model fitting Ξ and X, correspond to the best linear model fitting the centered data points $\tilde{\Xi} = \Xi - \frac{1_{N \times N}}{N}\Xi$ and centered embedded points $\tilde{X} = X - \frac{1_{N \times N}}{N}X$, which may be expressed as:

$$\min_{b} ||\Xi - XA - 1_{N \times 1}b||_2 = \left\| \tilde{\Xi} - \tilde{X}A \right\|_2$$

Indeed, the minimization of the left hand-side with respect to a row vector b of size $1 \times \delta$ gives:

$$\frac{\partial ||\Xi - XA - 1_{N \times 1}b||_2}{\partial b} = \sum_i \underbrace{\frac{\partial \left(\xi_i - x_i A - b\right)^2}{\partial b}}_{b - (\xi_i - x_i A)} = Nb - \sum_i \xi_i - x_i A$$

$$\frac{\partial ||\Xi - XA - 1_{N \times 1}b||_2}{\partial b} = 0 \Leftrightarrow b = \frac{1}{N}\sum_i \xi_i - x_i A \Leftrightarrow 1_{N \times 1}b = \frac{1_{N \times N}}{N}(\Xi - XA)$$

We may note that the best affine model always satisfies $\frac{1_{N \times N}}{N}\Xi = \left(\frac{1_{N \times N}}{N}X\right)A$

Furthermore, in PCA, the centering of X may be removed, since the minimization with respect to X and A always results in a centered X matrix. Indeed, considering the solution expressed as a projection, its mean is $\frac{1_{N \times N}}{N}X = \frac{1_{N \times N}}{N}(\tilde{\Xi}PP^\top) = 0$ (X is centered by design). In conclusion, PCA on the centered data matrix finds the best low-rank affine model approximated the data.

D.5   Link with Covariance and Gram Matrices

PCA may be interpreted in terms of covariance as mentioned in Sect. 5.1.1.1) or in terms of Gram matrix, which induces the equivalence with cMDS in a Euclidean space mentioned in Sect. 5.1.2.

Using the SVD decomposition of $\tilde{\Xi}$ presented above, we obtain the $\delta \times \delta$ covariance matrix $C_{\tilde{\Xi}}$ and the $N \times N$ centered Gram matrix $\Gamma_{\tilde{\Xi}}$:

$$C_{\tilde{\Xi}} = \tilde{\Xi}^\top \tilde{\Xi} = V \sqrt{\Lambda}^2 V^\top,$$

$$\Gamma_{\tilde{\Xi}} = \tilde{\Xi} \tilde{\Xi}^\top = U \sqrt{\Lambda}^2 U^\top.$$

The matrices U, V and $\sqrt{\Lambda}^2$ may thus be obtained by eigenvalue decomposition, either of $C_{\tilde{\Xi}}$ or of $\Gamma_{\tilde{\Xi}}$. Considering the equations $X = \tilde{\Xi}\mathring{V}$ and $X = \mathring{U}\sqrt{\mathring{\Lambda}}$, X may be computed based on the centered data matrix $\tilde{\Xi}$ and its associated covariance matrix $C_{\tilde{\Xi}}$ (providing $\mathring{V}$) or only based on the centered Gram matrix $\Gamma_{\tilde{\Xi}}$ (providing both $\mathring{U}$ and $\sqrt{\mathring{\Lambda}} = (\sqrt{\mathring{\Lambda}}^2)^{1/2}$). The second approach is used in cMDS to allow the use of non Euclidean metrics.

Moreover, we may show, that PCA is equivalent to finding the low rank representation best fitting the covariance and centered Gram matrices, which may be expressed as:

$$\mathrm{argmin}_{XA}||C_{\tilde{\Xi}} - C_{(XA)}||_2 = \mathrm{argmin}_{XA}||\Gamma_{\tilde{\Xi}} - \Gamma_{XA}||_2 = \mathrm{argmin}_{XA}||\tilde{\Xi} - XA||_2.$$

Indeed, using the orthogonal invariance of the Frobenius norm and the rank constraint of XA as previously, we get that the covariance and centered Gram matrices of XA must be:

$$\overbrace{\underbrace{(XA)^\top(XA)}_{C_{(XA)}}} = V\sqrt{\widehat{\Lambda}}^2 V^\top = V\sqrt{\widehat{\Lambda}}\overbrace{U^\top U}^{I_{r\times r}}\sqrt{\widehat{\Lambda}}V^\top,$$

$$\underbrace{\Gamma_{(XA)}}_{(XA)(XA)^\top} - U\sqrt{\widehat{\Lambda}}^2 U^\top = U\sqrt{\widehat{\Lambda}}\underbrace{V^\top V}_{I_{r\times r}}\sqrt{\widehat{\Lambda}}U^\top,$$

so that $XA = U\sqrt{\widehat{\Lambda}}V^\top$ is a solution of the minimization problem.

D.6 From Distances to Gram Matrix

We detail here the generalization of cMDS (considered as a Gram matrix formulation of PCA) to any set of metric data (see Sect. 5.1.2).

The centering formula for a Gram matrix is:

$$\boxed{\Gamma_{\tilde{\Xi}} = \left(I_{N\times N} - \frac{1_{N\times N}}{N}\right)\Gamma_{\Xi}\left(I_{N\times N} - \frac{1_{N\times N}}{N}\right)}$$

The centering effect may be justified by the following derivation:

$$\Gamma_{\tilde{\Xi}} = \tilde{\Xi}\tilde{\Xi}^{\mathsf{T}} = \left(\Xi - \frac{1_{N\times N}}{N}\Xi\right)\left(\Xi - \frac{1_{N\times N}}{N}\Xi\right)^{\mathsf{T}}$$

$$= \left(\Xi - \frac{1_{N\times N}}{N}\Xi\right)\left(\Xi^{\mathsf{T}} - \Xi^{\mathsf{T}}\frac{1_{N\times N}}{N}\right)$$

$$= \left(I_{N\times N} - \frac{1_{N\times N}}{N}\right)\underbrace{\Xi\Xi^{\mathsf{T}}}_{\Gamma_{\Xi}}\left(I_{N\times N} - \frac{1_{N\times N}}{N}\right)$$

or in the element-wise form:

$$\left\langle \xi_i - \frac{1}{N}\sum_l \xi_l, \xi_j - \frac{1}{N}\sum_m \xi_m \right\rangle =$$

$$\langle \xi_i, \xi_j \rangle - \frac{1}{N}\sum_m \langle \xi_i, \xi_m \rangle - \frac{1}{N}\sum_l \langle \xi_l, \xi_j \rangle + \frac{1}{N^2}\sum_l\sum_m \langle \xi_l, \xi_m \rangle.$$

Considering that any given distance matrix may be seen as the matrix of distances between point ξ_i in a Euclidean space [206], we have for each pair (i, j): $\Delta_{ij} = \sqrt{\langle \xi_i - \xi_j, \xi_i - \xi_j \rangle}$. This leads to (cosine law):

$$\Delta_{ij}^2 = \langle \xi_i, \xi_i \rangle - 2\langle \xi_i, \xi_j \rangle + \langle \xi_j, \xi_j \rangle. \tag{D.7}$$

This may be reformulated using matrix notations:

$$\Delta_{\Xi} \circ \Delta_{\Xi} = (\Gamma_{\Xi} \circ I_{N\times N})1_{N\times N} - 2\Gamma_{\Xi} + 1_{N\times N}(\Gamma_{\Xi} \circ I_{N\times N}),$$

with $\circ$ the Hadamard product (or term by term product), and thus $\Gamma_{\Xi} \circ I_{N\times N}$ is the diagonal matrix containing the squared norms of vectors $\mathrm{diag}(\dots, \langle \xi_i, \xi_i \rangle, \dots)$.

Application of the centering formula to Eq. (D.7) removes the terms $(\Gamma_{\Xi} \circ I_{N\times N})1_{N\times N}$ with constant rows and $1_{N\times N}(G_{\Xi} \circ I_{N\times N})$ with constant columns. Indeed:

$$\left(I_{N\times N} - \frac{1_{N\times N}}{N}\right)1_{N\times N} = 1_{N\times N}\left(I_{N\times N} - \frac{1_{N\times N}}{N}\right) = 0_{N\times N}.$$

As such, the centering of Eq. (D.7) gives the following formula for obtaining the centered Gram matrix from the distance matrix:

$$\boxed{\Gamma_{\tilde{\Xi}} = -\frac{1}{2}\left(I_{N\times N} - \frac{1_{N\times N}}{N}\right)\Delta_{\Xi} \circ \Delta_{\Xi}\left(I_{N\times N} - \frac{1_{N\times N}}{N}\right).}$$

D.6.1   Probabilistic Interpretation and Maximum Likelihood

A probabilistic interpretation of PCA models the data $\tilde{\Xi}$ with a latent variable model as detailed in Sect. 5.1.1.3. In that approach, the data points are in a subspace generated by the row vectors of A, parametrized by the latent variables of X, with an added isotropic Gaussian noise. Hence, the negative log likelihood, which must be minimized to find the most likely values of the latent variables $\{x_i\}$ may be expressed as:

$$-\log \mathscr{L}(X) = -\log \left(\prod_i \exp\left(\frac{-||\xi_i - x_i A||_2^2}{2\sigma^2} \right) \right) = \sum_i \frac{||\xi_i - x_i A||_2^2}{2\sigma^2},$$

which leads to:

$$-\log \mathscr{L} \propto ||\Xi - XA||_2^2.$$

Thus, minimizing the stress of PCA provides the most probable parameters of that model.

D.7   Nyström Approximation

The extension of spectral projection methods to out-of-sample data is detailed Sect. 7.3.2.1.

We consider a kernel Gram matrix $K_{\psi(\tilde{\Xi})}$ for a training set Ξ, with $\Psi :$ $\mathcal{D} \longrightarrow \mathcal{K}$ the basis expansion function (this may also be applied to cases without basis expansion). The definition of eigenvectors $\mathring{U}^\top$ associated to the d highest eigenvalues $\mathring{\Lambda} = \sqrt{\mathring{\Lambda}}^2$ as directions for which the transformation associated with the matrix $K_{\psi(\tilde{\Xi})}$ is locally a homothety, may be written with the following matrix formulation:

$$K_{\psi(\tilde{\Xi})} \mathring{U} = \mathring{U} \sqrt{\mathring{\Lambda}}^2.$$

Multiplying both sides of the equation by $\sqrt{\mathring{\Lambda}}^{-1}$ and using the embedding formula $X = \mathring{U}\sqrt{\mathring{\Lambda}}$ we get:

$$K_{\psi(\tilde{\Xi})} \underbrace{\mathring{U}\sqrt{\mathring{\Lambda}}}_{X} \sqrt{\mathring{\Lambda}}^{-2} = \underbrace{\mathring{U}\sqrt{\mathring{\Lambda}}}_{X}.$$

Reformulated point by point, this gives the required Nyström formula:

$$x_i = \sum_j \langle \Psi(\xi_i), \Psi(\xi_j) \rangle x_j \sqrt{\mathring{\Lambda}}^{-2}.$$

This may be used to extend the mapping to other points $\widehat{\xi}_i$, considering that they do not significantly impact the eigenvectors and eigenvalues if there is enough training points:

$$\widehat{x}_i = \sum_j \langle \Psi(\widehat{\xi}_i), \Psi(\xi_j) \rangle x_j \sqrt{\mathring{\Lambda}}^{-2}.$$

Conflict of Interest Statement

The Authors certify that they have **no** affiliations with or involvement in any organization or entity with any financial interest (such as honoraria; educational grants; participation in speakers' bureaus; membership, employment, consultancies, stock ownership, or other equity interest; and expert testimony or patent-licensing arrangements), or non-financial interest (such as personal or professional relationships, affiliations, knowledge or beliefs) in the subject matter or materials discussed in this manuscript.

Disclaimer Statement

The opinions expressed in this book are those of the Authors and do not necessarily reflect the views of their employers or any other affiliated organizations.

References

1. Aggarwal CC, Hinneburg A, Keim DA (2001) On the surprising behavior of distance metrics in high dimensional space. In: Van den Bussche J, Vianu V (eds) Database theory — ICDT 2001. Lecture notes in computer science. Springer, Berlin, pp 420–434
2. Akkucuk U, Carroll JD (2006) PARAMAP vs. Isomap: a comparison of two nonlinear mapping algorithms. J Classif 23(2):221–254
3. Allegra M, Facco E, Denti F, Laio A, Mira A (2020) Data segmentation based on the local intrinsic dimension. Sci Rep 10(1):1–12. https://doi.org/10.1038/s41598-020-72222-0
4. Alpaydin E, Kaynak C (1998) Cascading classifiers. Kybernetika 34(4):369–374
5. Amorim E, Brazil EV, Nonato LG, Samavati F, Sousa MC (2014) Multidimensional projection with radial basis function and control points selection. In: 2014 IEEE pacific visualization symposium, pp 209–216. https://doi.org/10.1109/PacificVis.2014.59
6. Amsaleg L, Chelly O, Furon T, Girard S, Houle ME, Kawarabayashi Ki, Nett M (2015) Estimating local intrinsic dimensionality. In: Proceedings of the 21th ACM SIGKDD international conference on knowledge discovery and data mining - KDD '15, ACM Press, Sydney, NSW, pp 29–38, https://doi.org/10.1145/2783258.2783405. http://dl.acm.org/citation.cfm?doid=2783258.2783405
7. Ankerst M, Berchtold S, Keim DA (1998) Similarity clustering of dimensions for an enhanced visualization of multidimensional data. In: Proceedings IEEE symposium on information visualization (Cat. No.98TB100258), pp 52–60. https://doi.org/10.1109/INFVIS.1998.729559
8. Asimov D (1985) The grand tour. SIAM J Sci Statist Comput. https://dl.acm.org/doi/abs/10.1137/0906011
9. Aupetit M (2006) Learning topology with the generative Gaussian graph and the EM algorithm. In: Weiss Y, Schölkopf B, Platt JC (eds) Advances in neural information processing systems, vol 18. MIT Press, Cambridge, pp 83–90. http://papers.nips.cc/paper/2922-learning-topology-with-the-generative-gaussian-graph-and-the-em-algorithm.pdf
10. Aupetit M (2007) Visualizing distortions and recovering topology in continuous projection techniques. Neurocomputing 70(7–9):1304–1330. https://doi.org/10.1016/j.neucom.2006.11.018. http://linkinghub.elsevier.com/retrieve/pii/S0925231206004814
11. Aupetit M, Sedlmair M (2016) SepMe: 2002 new visual separation measures. In: 2016 IEEE pacific visualization symposium (PacificVis), pp 1–8. https://doi.org/10.1109/PACIFICVIS.2016.7465244
12. Aupetit M, van der Maaten L (2015) Introduction to the special issue on visual analytics using multidimensional projections. Neurocomputing 150:543–545. https://doi.org/10.1016/j.neucom.2014.10.015. http://linkinghub.elsevier.com/retrieve/pii/S0925231214013265

13. Bejancu Jr A (1999) Local accuracy for radial basis function interpolation on finite uniform grids. J Approx Theory 99(2):242–257
14. Belkin M, Niyogi P (2002) Laplacian eigenmaps and spectral techniques for embedding and clustering. In: Advances in neural information processing systems, pp 585–591
15. Bellet A, Habrard A, Sebban M (2013) A survey on metric learning for feature vectors and structured data. arXiv:13066709 [cs, stat]. http://arxiv.org/abs/1306.6709
16. Bengio Y, Paiement JF, Vincent P, Delalleau O, Roux NL, Ouimet M (2004) Out-of-sample extensions for lle, isomap, mds, eigenmaps, and spectral clustering. In: Advances in neural information processing systems, pp 177–184
17. Bertin J (1983) Semiology of graphics. Madison, Wis
18. Bishop CM, James GD (1993) Analysis of multiphase flows using dual-energy gamma densitometry and neural networks. Nucl Instrum Methods Phys Res Section A: Accelerators, Spectrom Detectors Assoc Equipt 327(2):580–593. https://doi.org/10.1016/0168-9002(93)90728-Z. http://www.sciencedirect.com/science/article/pii/016890029390728Z
19. Blondel VD, Guillaume JL, Lambiotte R, Lefebvre E (2008) Fast unfolding of communities in large networks. J Statist Mech Theory Exp 2008(10):P10008. https://doi.org/10.1088/1742-5468/2008/10/P10008. https://doi.org/10.1088%2F1742-5468%2F2008%2F10%2Fp10008
20. Böhm JN, Berens P, Kobak D (2020) A unifying perspective on neighbor embeddings along the attraction-repulsion spectrum. arXiv:200708902 [cs, stat]. http://arxiv.org/abs/2007.08902
21. Borg I, Groenen PJF (2005) Modern multidimensional scaling: theory and applications. Springer series in statistics, 2nd edn. Springer, New York
22. Borland D, Ii RMT (2007) Rainbow color map (still) considered harmful. IEEE Comput Graphics Appl 27(2):14–17. https://doi.org/10.1109/MCG.2007.323435
23. Börner K (2015) Atlas of knowledge: Anyone can map. MIT Press, Cambridge
24. Bregman LM (1967) The relaxation method of finding the common point of convex sets and its application to the solution of problems in convex programming. USSR Comput Math Math Phys 7(3):200–217. https://doi.org/10.1016/0041-5553(67)90040-7. http://www.sciencedirect.com/science/article/pii/0041555367900407
25. Breiman L (2001) Random forests. Mach Learn 45(1):5–32
26. Breunig MM, Kriegel HP, Ng RT, Sander J (2000) LOF: identifying density-based local outliers. In: Proceedings of the 2000 ACM SIGMOD international conference on management of data, pp 93–104
27. Brewer C (2021) ColorBrewer: Color advice for maps. http://colorbrewer2.org/. Accessed 02 Feb 2019
28. Burges CJC (2009) Dimension Reduction: a guided tour. Found Trends® Mach Learn 2(4):275–364. https://doi.org/10.1561/2200000002. http://www.nowpublishers.com/article/Details/MAL-002
29. Camastra F, Vinciarelli A (2002) Estimating the intrinsic dimension of data with a fractal-based method. IEEE Trans Pattern Analy Mach Intell 24(10):1404–1407. https://doi.org/10.1109/TPAMI.2002.1039212
30. Campello RJ, Moulavi D, Sander J (2013) Density-based clustering based on hierarchical density estimates. In: Pacific-Asia conference on knowledge discovery and data mining. Springer, Berlin, pp 160–172
31. Carreira-Perpinán MA (2010) The elastic embedding algorithm for dimensionality reduction. In: Proceedings of the 27th international conference on international conference on machine learning, vol 10, pp 167–174
32. Cattell RB (1966) The scree test for the number of factors. Multivar Behav Res 1(2):245–276. https://doi.org/10.1207/s15327906mbr0102_10
33. Chen L, Buja A (2009) Local multidimensional scaling for nonlinear dimension reduction, graph drawing, and proximity analysis. J Am Statist Assoc 104(485):209–219. https://doi.org/10.1198/jasa.2009.0111. https://amstat.tandfonline.com/doi/abs/10.1198/jasa.2009.0111

34. Coifman RR, Lafon S, Lee AB, Maggioni M, Nadler B, Warner F, Zucker SW (2005) Geometric diffusions as a tool for harmonic analysis and structure definition of data: diffusion maps. Procee Natl Acad Sci 102(21):7426–7431. https://doi.org/10.1073/pnas.0500334102. https://www.pnas.org/content/102/21/7426

35. Colange B, Vuillon L, Lespinats S, Dutykh D (2019) Interpreting distortions in dimensionality reduction by superimposing neighbourhood graphs. In: 2019 IEEE visualization conference (VIS). IEEE, Piscatway, pp 211–215

36. Colange B, Peltonen J, Aupetit M, Dutykh D, Lespinats S (2020) Steering distortions to preserve classes and neighbors in supervised dimensionality reduction. In: Larochelle H, Ranzato M, Hadsell R, Balcan MF, Lin H (eds) Advances in neural information processing systems, vol 33. Curran Associates, Red Hook, pp 13214–13225. https://proceedings.neurips.cc/paper/2020/file/99607461cdb9c26e2bd5f31b12dcf27a-Paper.pdf

37. Collins M, Dasgupta S, Schapire RE (2002) A generalization of principal components analysis to the exponential family. In: Dietterich TG, Becker S, Ghahramani Z (eds) Advances in neural information processing systems, vol 14. MIT Press, Cambridge, pp 617–624. http://papers.nips.cc/paper/2078-a-generalization-of-principal-components-analysis-to-the-exponential-family.pdf

38. Comaniciu D, Meer P (2002) Mean shift: a robust approach toward feature space analysis. IEEE Trans Pattern Analy Mach Intell 24(5):603–619. https://doi.org/10.1109/34.1000236

39. Commission Internationale de l'Eclairage (1978) Recommendations on uniform color spaces, color-difference equations, psychometric color terms. CIE, Paris

40. Cook D, Buja A, Cabrera J, Hurley C (1995) Grand tour and projection pursuit. J Computat Graph Statist 4(3):155–172

41. Cook J, Sutskever I, Mnih A, Hinton G (2007) Visualizing similarity data with a mixture of maps. In: Artificial intelligence and statistics, pp 67–74

42. Cortes C, Vapnik V (1995) Support-vector networks. Mach Learn 20(3):273–297. https://doi.org/10.1007/BF00994018

43. Cover T, Hart P (1967) Nearest neighbor pattern classification. IEEE Trans Inf Theory 13(1):21–27. https://doi.org/10.1109/TIT.1967.1053964

44. Cutura R, Aupetit M, Fekete JD, Sedlmair M (2020) Comparing and exploring high-dimensional data with dimensionality reduction algorithms and matrix visualizations. In: AVI'20-international conference on advanced visual interfaces. https://doi.org/10.1145/3399715.3399875. https://hal.inria.fr/hal-02861899

45. Dasgupta S, Freund Y (2008) Random projection trees and low dimensional manifolds. In: Proceedings of the fortieth annual ACM symposium on theory of computing, pp 537–546

46. Dayhoff M, Schwartz R, Orcutt B (1978) 22 a model of evolutionary change in proteins. Atlas Protein Seq Struct 5:345–352

47. de Bodt C, Mulders D, Verleysen M, Lee JA (2018) Perplexity-free t-SNE and twice Student tt-SNE. In: ESANN. https://dial.uclouvain.be/pr/boreal/object/boreal:200844

48. de Bodt C, Mulders D, Sánchez DL, Verleysen M, Lee JA (2019) Class-aware t-SNE: cat-SNE. In: ESANN

49. De Marchi S (2013) Four lectures on radial basis functions

50. de Ridder D, Kouropteva O, Okun O, Pietikäinen M, Duin RP (2003) Supervised locally linear embedding. In: Artificial neural networks and neural information processing–ICANN/ICONIP 2003. Springer, Berlin, pp 333–341

51. de Ridder D, Loog M, Reinders MJT (2004) Local fisher embedding. In: Proceedings of the 17th international conference on pattern recognition, 2004. ICPR 2004, vol 2. pp 295–298. https://doi.org/10.1109/ICPR.2004.1334176

52. De Silva V, Tenenbaum JB (2004) Sparse multidimensional scaling using landmark points. Technical Report, technical report, Stanford University

53. Degret F, Lespinats S (2018) Circular background decreases misunderstanding of multidimensional scaling results for naive readers. In: MATEC Web of conferences. EDP sciences, vol 189, p 10002

54. Demartines P, Hérault J (1997) Curvilinear component analysis: a self-organizing neural network for nonlinear mapping of data sets. IEEE Trans Neural Netw 8(1):148–154
55. Diaconis P, Shahshahani M (1987) The subgroup algorithm for generating uniform random variables. Probab Eng Inf Sci 1(01):15. https://doi.org/10.1017/S0269964800000255. http://www.journals.cambridge.org/abstract_S0269964800000255
56. Dong W, Moses C, Li K (2011) Efficient k-nearest neighbor graph construction for generic similarity measures. In: Proceedings of the 20th international conference on World wide web, pp 577–586
57. Donoho DL (2000) High-dimensional data analysis: the curses and blessings of dimensionality. AMS Math Challenges Lecture 1(32):375
58. Dua D, Karra Taniskidou E (2017) UCI machine learning repository. http://archive.ics.uci.edu/ml
59. Eckmann JP, Ruelle D (1992) Fundamental limitations for estimating dimensions and Lyapunov exponents in dynamical systems. Physica D Nonlinear Phenom 56(2–3):185–187
60. Espadoto M, Martins RM, Kerren A, Hirata NST, Telea AC (2019) Towards a Quantitative survey of dimension reduction techniques. IEEE Trans Visualiz Comput Graph 27:1–1. https://doi.org/10.1109/TVCG.2019.2944182
61. Ester M, Kriegel HP, Sander J, Xu X, et al (1996) A density-based algorithm for discovering clusters in large spatial databases with noise. In: Kdd, vol 96, pp 226–231
62. Facco E, d'Errico M, Rodriguez A, Laio A (2017) Estimating the intrinsic dimension of datasets by a minimal neighborhood information. Sci Rep 7(1):12140. https://doi.org/10.1038/s41598-017-11873-y. https://www.nature.com/articles/s41598-017-11873-y
63. Fanty M, Cole R (1991) Spoken letter recognition. In: Advances in neural information processing systems, pp 220–226
64. Faure G (2018) Etude de défauts critiques des installations solaires thermiques de grande dimension: définition, modélisation et diagnostic. PhD thesis, Université Grenoble Alpes
65. Faust R, Glickenstein D, Scheidegger C (2019) DimReader: axis lines that explain non-linear projections. IEEE Trans Vis Comput Graph 25(1):481–490. https://doi.org/10.1109/TVCG.2018.2865194
66. Felsenstein J (1993) PHYLIP (phylogeny inference package), version 3.5 c. Joseph Felsenstein
67. Fisher RA (1936) The use of multiple measurements in taxonomic problems. Ann Eugenics 7(2):179–188. https://doi.org/10.1111/j.1469-1809.1936.tb02137.x. https://onlinelibrary.wiley.com/doi/abs/10.1111/j.1469-1809.1936.tb02137.x
68. Fornberg B, Flyer N (2005) Accuracy of radial basis function interpolation and derivative approximations on 1-D infinite grids. Adv Computat Math 23(1–2):5–20
69. Friedman JH, Bentley JL, Finkel RA (1977) An algorithm for finding best matches in logarithmic expected time. ACM Trans Math Softw (TOMS) 3(3):209–226
70. Fruchterman TMJ, Reingold EM (1991) Graph drawing by force-directed placement. Softw Practice Exper 21(11):1129–1164. https://doi.org/10.1002/spe.4380211102. https://onlinelibrary.wiley.com/doi/abs/10.1002/spe.4380211102
71. Fujiwara T, Chou J, Shilpika S, Xu P, Ren L, Ma K (2020) An incremental dimensionality reduction method for visualizing streaming multidimensional data. IEEE Trans Vis Comput Graph 26(1):418–428. https://doi.org/10.1109/TVCG.2019.2934433
72. Gabriel KR (1971) The biplot graphic display of matrices with application to principal component analysis. Biometrika 58(3):453–467
73. Gaillard P, Aupetit M, Govaert G (2008) Learning topology of a labeled data set with the supervised generative Gaussian graph. Neurocomput 71(7):1283–1299. https://doi.org/10.1016/j.neucom.2007.12.028. http://www.sciencedirect.com/science/article/pii/S0925231208000635
74. Geissbuehler M, Lasser T (2013) How to display data by color schemes compatible with red-green color perception deficiencies. Optics Express 21(8):9862–9874. https://doi.org/10.1364/OE.21.009862. https://www.osapublishing.org/oe/abstract.cfm?uri=oe-21-8-9862

75. Geng X, Zhan D-C, Zhou Z-H (2005) Supervised nonlinear dimensionality reduction for visualization and classification. IEEE Trans Syst Man Cyb Part B (Cybernetics) 35(6):1098–1107. https://doi.org/10.1109/TSMCB.2005.850151

76. Geoffroy H (2020) Méthode pour la détection de défauts des systèmes énergétiques - Couplage expertise et méthode de réduction de dimensions. PhD Thesis, Université Savoie Mont-Blanc

77. Geoffroy H, Berger J, Colange B, Lespinats S, Dutykh D, Buhé C, Sauce G (2019) Use of multidimensional scaling for fault detection or monitoring support in a continuous commissioning. In: Building simulation. Springer, Berlin

78. Gisbrecht A, Schulz A, Hammer B (2015) Parametric nonlinear dimensionality reduction using kernel t-SNE. Neurocomputing 147:71–82. https://doi.org/10.1016/j.neucom.2013.11.045. http://www.sciencedirect.com/science/article/pii/S0925231214007036

79. Goldberger J, Hinton GE, Roweis ST, Salakhutdinov RR (2005) Neighbourhood Components Analysis. In: Saul LK, Weiss Y, Bottou L (eds) Advances in neural information processing systems, vol 17. MIT Press, Cambridge, pp 513–520. http://papers.nips.cc/paper/2566-neighbourhood-components-analysis.pdf

80. Gower JC (1966) Some distance properties of latent root and vector methods used in multivariate analysis. Biometrika 53(3-4):325–338

81. Granata D, Carnevale V (2016) Accurate estimation of the intrinsic dimension using graph distances: unraveling the geometric complexity of datasets. Sci Rep 6:31377. https://doi.org/10.1038/srep31377. https://www.nature.com/articles/srep31377

82. Grassberger P, Procaccia I (1983) Measuring the strangeness of strange attractors. Physica D Nonlinear Phenomena 9(1):189–208. https://doi.org/10.1016/0167-2789(83)90298-1. http://www.sciencedirect.com/science/article/pii/0167278983902981

83. Guillet N, Primot C, Degret F, Thivel PX (2017) In-operando techniques for battery monitoring and safety issues prevention. In: European battery, hybrid and fuel cell electric vehicle congress

84. Ham J, Lee DD, Mika S, Schölkopf B (2004) A kernel view of the dimensionality reduction of manifolds. In: Proceedings of the twenty-first international conference on machine learning, p 47

85. Hastie T, Tibshirani R, Friedman J (2009) The elements of statistical learning. Springer series in statistics, vol 1. Springer, New York

86. Heulot N, Aupetit M, Fekete JD (2013) Proxilens: Interactive exploration of high-dimensional data using projections. In: VAMP: EuroVis workshop on visual analytics using multidimensional projections. The Eurographics Association, Aire-la-Ville

87. Hinton GE, Roweis ST (2003) Stochastic neighbor embedding. In: Advances in neural information processing systems, pp 857–864

88. Hinton GE, Salakhutdinov RR (2006) Reducing the dimensionality of data with neural networks. Science 313(5786):504–507. https://doi.org/10.1126/science.1127647. https://science.sciencemag.org/content/313/5786/504

89. Holten D, Wijk JJV (2009) Force-directed edge bundling for graph visualization. Comput Graph Forum 28(3):983–990. https://doi.org/10.1111/j.1467-8659.2009.01450.x. https://onlinelibrary.wiley.com/doi/abs/10.1111/j.1467-8659.2009.01450.x

90. Hunter JD (2007) Matplotlib: A 2d graphics environment. Comput Sci Eng 9(3):90–95. https://doi.org/10.1109/MCSE.2007.55

91. Hurter C, Ersoy O, Telea A (2012) Graph bundling by Kernel density estimation. Comput Graph Forum 31(3pt1):865–874. https://doi.org/10.1111/j.1467-8659.2012.03079.x. https://onlinelibrary.wiley.com/doi/abs/10.1111/j.1467-8659.2012.03079.x

92. Inselberg A, Dimsdale B (1990) Parallel coordinates: a tool for visualizing multi-dimensional geometry. In: Proceedings of the first IEEE conference on visualization: visualization '90, pp 361–378. https://doi.org/10.1109/VISUAL.1990.146402

93. Joia P, Coimbra D, Cuminato JA, Paulovich FV, Nonato LG (2011) Local affine multidimensional projection. IEEE Trans Vis Comput Graph 17(12):2563–2571. https://doi.org/10.1109/TVCG.2011.220

94. Jolliffe IT (1986) Principal components in regression analysis. In: Jolliffe IT (ed) Principal component analysis. Springer series in statistics. Springer, New York, pp 129–155. https://doi.org/10.1007/978-1-4757-1904-8_8

95. Kammer D, Keck M, Gründer T, Maasch A, Thom T, Kleinsteuber M, Groh R (2020) Glyphboard: visual exploration of high-dimensional data combining glyphs with dimensionality reduction. IEEE Trans Vis Comput Graph 26(4):1661–1671. https://doi.org/10.1109/TVCG.2020.2969060

96. Kaufmann L (1987) Clustering by means of medoids. In: Proceedings of the statistical data analysis based on the L1 norm conference, Neuchatel, 1987, pp 405–416

97. Kitazono J, Grozavu N, Rogovschi N, Omori T, Ozawa S (2016) t-distributed stochastic neighbor embedding with inhomogeneous degrees of freedom. In: Hirose A, Ozawa S, Doya K, Ikeda K, Lee M, Liu D (eds) Neural information processing. Lecture notes in computer science. Springer International Publishing, Cham, pp 119–128

98. Klambauer G, Unterthiner T, Mayr A, Hochreiter S (2017) Self-normalizing neural networks. In: Guyon I, Luxburg UV, Bengio S, Wallach H, Fergus R, Vishwanathan S, Garnett R (eds) Advances in neural information processing systems, vol 30. Curran Associates, Red Hook, pp 971–980. http://papers.nips.cc/paper/6698-self-normalizing-neural-networks.pdf

99. Kruskal JB (1964) Multidimensional scaling by optimizing goodness of fit to a nonmetric hypothesis. Psychometrika 29(1):1–27. https://doi.org/10.1007/BF02289565

100. Kruskal JB (1964) Nonmetric multidimensional scaling: a numerical method. Psychometrika 29(2):115–129

101. Landesberger TV, Kuijper A, Schreck T, Kohlhammer J, Wijk JJv, Fekete JD, Fellner DW (2011) Visual analysis of large graphs: state-of-the-art and future research challenges. Comput Graph Forum 30(6):1719–1749. https://doi.org/10.1111/j.1467-8659.2011.01898.x. https://onlinelibrary.wiley.com/doi/abs/10.1111/j.1467-8659.2011.01898.x

102. LeCun Y, Bengio Y, et al (1995) Convolutional networks for images, speech, and time series.Handbook Brain Theory Neural Netw 3361(10):1995

103. Lecun Y, Bottou L, Bengio Y, Haffner P (1998) Gradient-based learning applied to document recognition. Procee IEEE 86(11):2278–2324. https://doi.org/10.1109/5.726791

104. Lee DD, Seung HS (1999) Learning the parts of objects by non-negative matrix factorization. Nature 401(6755):788–791. https://doi.org/10.1038/44565. https://www.nature.com/articles/44565

105. Lee JA, Verleysen M (2009) Quality assessment of dimensionality reduction: Rank-based criteria. Neurocomputing 72(7):1431–1443. https://doi.org/10.1016/j.neucom.2008.12.017. http://www.sciencedirect.com/science/article/pii/S0925231209000101

106. Lee JA, Verleysen M (2010) Scale-independent quality criteria for dimensionality reduction. Pattern Recogn Lett 31(14):2248–2257. https://doi.org/10.1016/j.patrec.2010.04.013. http://www.sciencedirect.com/science/article/pii/S0167865510001364

107. Lee JA, Verleysen M (2014) Two key properties of dimensionality reduction methods. In: 2014 IEEE symposium on computational intelligence and data mining (CIDM). IEEE, Piscataway, pp 163–170

108. Lee JA, Lendasse A, Verleysen M (2004) Nonlinear projection with curvilinear distances: isomap versus curvilinear distance analysis. Neurocomputing 57:49–76. https://doi.org/10.1016/j.neucom.2004.01.007. https://linkinghub.elsevier.com/retrieve/pii/S0925231204000645

109. Lee JA, Renard E, Bernard G, Dupont P, Verleysen M (2013) Type 1 and 2 mixtures of Kullback–Leibler divergences as cost functions in dimensionality reduction based on similarity preservation. Neurocomputing 112:92–108. https://doi.org/10.1016/j.neucom.2012.12.036. http://linkinghub.elsevier.com/retrieve/pii/S0925231213001471

110. Lee JA, Peluffo-Ordóñez DH, Verleysen M (2015) Multi-scale similarities in stochastic neighbour embedding: Reducing dimensionality while preserving both local and global structure. Neurocomputing 169:246–261. https://doi.org/10.1016/j.neucom.2014.12.095. https://linkinghub.elsevier.com/retrieve/pii/S0925231215003641

111. Lespinats S (2018) Method for determining the state of a system, method for determining an optimal projection method and device implementing said methods

112. Lespinats S, Aupetit M (2011) CheckViz: Sanity check and topological clues for linear and non-linear mappings. Comput Graph Forum 30(1):113–125. https://doi.org/10.1111/j.1467-8659.2010.01835.x. http://doi.wiley.com/10.1111/j.1467-8659.2010.01835.x

113. Lespinats S, Fertil B (2011) Colorphylo: a color code to accurately display taxonomic classifications. Evolut Bioinf 7:EBO–S7565

114. Lespinats S, Verleysen M, Giron A, Fertil B (2007) DD-HDS: a method for visualization and exploration of high-dimensional data. IEEE Trans Neural Netw 18(5):1265–1279. https://doi.org/10.1109/TNN.2007.891682

115. Lespinats S, Fertil B, Villemain P, Hérault J (2009) RankVisu: mapping from the neighborhood network. Neurocomputing 72(13):2964–2978. https://doi.org/10.1016/j.neucom.2009.04.008. http://www.sciencedirect.com/science/article/pii/S0925231209001544

116. Lespinats S, Aupetit M, Meyer-Base A (2015) ClassiMap: a new dimension reduction technique for exploratory data analysis of labeled data. Int J Pattern Recogn Artif Intell 29:150505235857008. https://doi.org/10.1142/S0218001415510088

117. Lespinats S, De Clerck O, Colange B, Gorelova V, Grando D, Maréchal E, Van Der Straeten D, Rébeillé F, Bastien O (2019) Phylogeny and sequence space: a combined approach to analyze the evolutionary trajectories of homologous proteins. the case study of aminodeoxychorismate synthase. Acta Biotheoretica 68:139–156. https://doi.org/10.1007/s10441-019-09352-0. https://doi.org/10.1007/s10441-019-09352-0

118. Lloyd S (1982) Least squares quantization in PCM. IEEE Trans Inf Theory 28(2):129–137. https://doi.org/10.1109/TIT.1982.1056489

119. Lu Y, Lu J (2020) A universal approximation theorem of deep neural networks for expressing probability distributions. In: Larochelle H, Ranzato M, Hadsell R, Balcan MF, Lin H (eds) Advances in neural information processing systems, , vol 33. Curran Associates, Red Hook, pp 3094–3105. https://proceedings.neurips.cc/paper/2020/file/2000f6325dfc4fc3201fc45ed01c7a5d-Paper.pdf

120. Lu M, Wang S, Lanir J, Fish N, Yue Y, Cohen-Or D, Huang H (2020) Winglets: visualizing association with uncertainty in multi-class scatterplots. IEEE Trans Vis Comput Graph 26(1):770–779. https://doi.org/10.1109/TVCG.2019.2934811

121. Mandelbrot B (1967) How long is the coast of britain? statistical self-similarity and fractional dimension. Science 156(3775):636–638. https://doi.org/10.1126/science.156.3775.636. https://science.sciencemag.org/content/156/3775/636

122. Martins RM, Coimbra DB, Minghim R, Telea A (2014) Visual analysis of dimensionality reduction quality for parameterized projections. Comput Graph 41:26–42. https://doi.org/10.1016/j.cag.2014.01.006. https://linkinghub.elsevier.com/retrieve/pii/S0097849314000235

123. Martins RM, Minghim R, Telea AC (2015) Explaining neighborhood preservation for multidimensional projections. In: CGVC, Eurographics - European association for computer graphics, pp 7–14. http://urn.kb.se/resolve?urn=urn:nbn:se:lnu:diva-73243

124. Mayorga A, Gleicher M (2013) Splatterplots: overcoming overdraw in scatter plots. IEEE Trans Vis Comput Graph 19(9):1526–1538. https://doi.org/10.1109/TVCG.2013.65

125. McInnes L, Healy J, Melville J (2018) UMAP: uniform manifold approximation and projection for dimension reduction. arXiv:180203426 [cs, stat]. http://arxiv.org/abs/1802.03426

126. McLachlan GJ, Peel D (2000) Finite mixture models. Wiley, New York

127. Meyer GW, Greenberg DP (1980) Perceptual color spaces for computer graphics. ACM SIGGRAPH Comput Graph 14:254–261. https://dl.acm.org/doi/abs/10.1145/965105.807502

128. Mika S, Ratsch G, Weston J, Scholkopf B, Mullers KR (1999) Fisher discriminant analysis with kernels. In: Neural networks for signal processing IX: proceedings of the 1999 IEEE signal processing society workshop (Cat. No.98TH8468), pp 41–48. https://doi.org/10.1109/NNSP.1999.788121

129. Mokbel B, Lueks W, Gisbrecht A, Hammer B (2013) Visualizing the quality of dimensionality reduction. Neurocomputing 112:109–123. https://doi.org/10.1016/j.neucom.2012.11.046. http://www.sciencedirect.com/science/article/pii/S0925231213002439

130. Moon KR, van Dijk D, Wang Z, Gigante S, Burkhardt DB, Chen WS, Yim K, Elzen Avd, Hirn MJ, Coifman RR, Ivanova NB, Wolf G, Krishnaswamy S (2019) Visualizing structure and transitions in high-dimensional biological data. Nat Biotechnol 37(12):1482–1492. https://doi.org/10.1038/s41587-019-0336-3. https://www.nature.com/articles/s41587-019-0336-3

131. Morrison A, Ross G, Chalmers M (2003) Fast multidimensional scaling through sampling, springs and interpolation. Inf Vis 2(1):68–77

132. Morrow B, Manz T, Chung AE, Gehlenborg N, Gotz D (2019) Periphery plots for contextualizing heterogeneous time-based charts. In: 2019 IEEE visualization conference (VIS), pp 1–5. https://doi.org/10.1109/VISUAL.2019.8933582

133. Motta R, Minghim R, de Andrade Lopes A, F Oliveira MC (2015) Graph-based measures to assist user assessment of multidimensional projections. Neurocomputing 150:583–598. https://doi.org/10.1016/j.neucom.2014.09.063. https://linkinghub.elsevier.com/retrieve/pii/S0925231214012909

134. Müllner D (2013) fastcluster: fast hierarchical, agglomerative clustering routines for R and Python. J Statist Softw 53(9):1–18

135. Nadaraya EA (1964) On estimating regression. Theory Probab Appl 9(1):141–142. https://doi.org/10.1137/1109020. https://epubs.siam.org/doi/abs/10.1137/1109020

136. Nene SA, Nayar SK, Murase H, et al. (1996) Columbia object image library (coil-100). COIL-100 Technical report 6

137. Newman MEJ (2005) Power laws, Pareto distributions and Zipf's law. Contemporary Phys 46(5):323–351. https://doi.org/10.1080/00107510500052444

138. Nocedal J, Wright SJ (1999) Numerical optimization. Springer series in operations research. Springer, New York

139. Nonato LG, Aupetit M (2019) Multidimensional projection for visual analytics: linking techniques with distortions, tasks, and layout enrichment. IEEE Trans Vis Comput Graph 25(8):2650–2673. https://doi.org/10.1109/TVCG.2018.2846735

140. Paulovich FV, Silva CT, Nonato LG (2010) Two-phase mapping for projecting massive data sets. IEEE Trans Vis Comput Graph 16(6):1281–1290. https://doi.org/10.1109/TVCG.2010.207

141. Pearson K (1901) LIII. On lines and planes of closest fit to systems of points in space. London Edinburgh Dublin Philos Mag J Sci 2(11):559–572. https://doi.org/10.1080/14786440109462720. https://www.tandfonline.com/doi/ref/10.1080/14786440109462720

142. Pedregosa F, Varoquaux G, Gramfort A, Michel V, Thirion B, Grisel O, Blondel M, Prettenhofer P, Weiss R, Dubourg V, Vanderplas J, Passos A, Cournapeau D, Brucher M, Perrot M, Duchesnay E (2011) Scikit-learn: machine learning in Python. J Mach Learn Res 12:2825–2830

143. Pekalska E, de Ridder D, Duin RP, Kraaijveld MA (1999) A new method of generalizing sammon mapping with application to algorithm speed-up. In: ASCI, vol 99, pp 221–228

144. Peltonen J, Klami A, Kaski S (2004) Improved learning of Riemannian metrics for exploratory analysis. Neural Netw 17(8):1087–1100. https://doi.org/10.1016/j.neunet.2004.06.008. http://www.sciencedirect.com/science/article/pii/S0893608004001558

145. Peltonen J, Aidos H, Kaski S (2009) Supervised nonlinear dimensionality reduction by Neighbor Retrieval. In: 2009 IEEE international conference on acoustics, speech and signal processing, pp 1809–1812. https://doi.org/10.1109/ICASSP.2009.4959957

146. Peysakhovich V, Hurter C, Telea A (2015) Attribute-driven edge bundling for general graphs with applications in trail analysis. In: 2015 IEEE Pacific visualization symposium (PacificVis), pp 39–46. https://doi.org/10.1109/PACIFICVIS.2015.7156354, ISSN: 2165-8773

147. Pezzotti N, Thijssen J, Mordvintsev A, Höllt T, Lew BV, Lelieveldt BPF, Eisemann E, Vilanova A (2020) GPGPU linear complexity t-SNE optimization. IEEE Trans Vis Comput Graph 26(1):1172–1181. https://doi.org/10.1109/TVCG.2019.2934307

148. Powell MJ (2005) Five lectures on radial basis functions
149. Qiu H, Hancock ER (2007) Clustering and embedding using commute times. IEEE Trans Pattern Analy Mach Intell 29(11):1873–1890. https://doi.org/10.1109/TPAMI.2007.1103
150. Rathore R, Leggon Z, Lessard L, Schloss KB (2020) Estimating color-concept associations from image statistics. IEEE Trans Vis Comput Graph 26(1):1226–1235. https://doi.org/10.1109/TVCG.2019.2934536
151. Rhodes JS, Cutler A, Wolf G, Moon KR (2020) Supervised visualization for data exploration. Preprint arXiv:200608701. https://arxiv.org/abs/2006.08701v1
152. Rogowitz BE, Treinish LA (1998) Data visualization: the end of the rainbow. IEEE Spectrum 35(12):52–59. https://doi.org/10.1109/6.736450
153. Rougier NP, Droettboom M, Bourne PE (2014) Ten simple rules for better figures. PLOS Computat Biol 10(9):e1003833. https://doi.org/10.1371/journal.pcbi.1003833. https://journals.plos.org/ploscompbiol/article?id=10.1371/journal.pcbi.1003833
154. Roweis ST, Saul LK (2000) Nonlinear dimensionality reduction by locally linear embedding. Science 290(5500):2323–2326. https://doi.org/10.1126/science.290.5500.2323. http://science.sciencemag.org/content/290/5500/2323
155. Sacha D, Zhang L, Sedlmair M, Lee JA, Peltonen J, Weiskopf D, North SC, Keim DA (2017) Visual interaction with dimensionality reduction: A structured literature analysis. IEEE Trans Vis Comput Graph 23(1):241–250
156. Sakoe H, Chiba S (1978) Dynamic programming algorithm optimization for spoken word recognition. IEEE Trans Acoustics Speech Signal Proc 26(1):43–49. https://doi.org/10.1109/TASSP.1978.1163055
157. Salakhutdinov R, Hinton G (2007) Learning a nonlinear embedding by preserving class neighbourhood structure. In: Artificial intelligence and statistics, pp 412–419
158. Sammon JW (1969) A nonlinear mapping for data structure analysis. IEEE Trans Comput C-18(5):401–409. https://doi.org/10.1109/T-C.1969.222678
159. Schölkopf B, Smola A, Müller KR (1998) Nonlinear component analysis as a Kernel eigenvalue problem. Neural Comput 10(5):1299–1319. https://doi.org/10.1162/089976698300017467
160. Schreck T, von Landesberger T, Bremm S (2010) Techniques for precision-based visual analysis of projected data. Inf Vis 9(3):181–193. https://doi.org/10.1057/ivs.2010.2
161. Schubert E, Gertz M (2017) Intrinsic t-stochastic neighbor embedding for visualization and outlier detection. In: Beecks C, Borutta F, Kröger P, Seidl T (eds) Similarity search and applications, vol 10609. Springer International Publishing, Cham, pp 188–203. https://doi.org/10.1007/978-3-319-68474-1_13. http://link.springer.com/10.1007/978-3-319-68474-1_13
162. Sedlmair M, Aupetit M (2015) Data-driven evaluation of visual quality measures. Comput Graph Forum 34(3):201–210. https://doi.org/10.1111/cgf.12632. https://onlinelibrary.wiley.com/doi/abs/10.1111/cgf.12632
163. Sedlmair M, Munzner T, Tory M (2013) Empirical guidance on scatterplot and dimension reduction technique choices. IEEE Trans Vis Comput Graph 19(12):2634–2643. https://doi.org/10.1109/TVCG.2013.153
164. Seifert C, Sabol V, Kienreich W (2010) Stress maps: analysing local phenomena in dimensionality reduction based visualisations. In: Proceedings of the 1st European symposium on visual analytics science and technology (EuroVAST'10), Bordeaux, vol 1
165. Selassie D, Heller B, Heer J (2011) Divided edge bundling for directional network data. IEEE Trans Vis Comput Graph 17(12):2354–2363. https://doi.org/10.1109/TVCG.2011.190
166. Shepard RN (1962) The analysis of proximities: multidimensional scaling with an unknown distance function. II. Psychometrika 27(3):219–246. https://doi.org/10.1007/BF02289621. https://doi.org/10.1007/BF02289621
167. Sibson R (1981) A brief description of natural neighbour interpolation. In: Interpreting multivariate data. Wiley, New York
168. Sips M, Neubert B, Lewis JP, Hanrahan P (2009) Selecting good views of high-dimensional data using class consistency. Comput Graph Forum 28(3):831–838. https://

doi.org/10.1111/j.1467-8659.2009.01467.x. https://onlinelibrary.wiley.com/doi/abs/10.1111/j.1467-8659.2009.01467.x

169. Smart S, Wu K, Szafir DA (2020) Color crafting: automating the construction of designer quality color ramps. IEEE Trans Vis Comput Graph 26(1):1215–1225. https://doi.org/10.1109/TVCG.2019.2934284

170. Stahnke J, Dörk M, Müller B, Thom A (2016) Probing projections: interaction techniques for interpreting arrangements and errors of dimensionality reductions. IEEE Trans Vis Comput Graph 22(1):629–638. https://doi.org/10.1109/TVCG.2015.2467717

171. Sudderth E (2012) Introduction to machine learning. http://cs.brown.edu/courses/cs195-5/spring2012/lectures/2012-01-26_overview.pdf

172. Sugiyama M (2006) Local fisher discriminant analysis for supervised dimensionality reduction. In: Proceedings of the 23rd international conference on machine learning, pp 905–912

173. Szubert B, Cole JE, Monaco C, Drozdov I (2019) Structure-preserving visualisation of high dimensional single-cell datasets. Sci Rep 9(1):1–10. https://doi.org/10.1038/s41598-019-45301-0. https://www.nature.com/articles/s41598-019-45301-0

174. Tang J, Liu J, Zhang M, Mei Q (2016) Visualizing large-scale and high-dimensional data. In: Proceedings of the 25th international conference on world wide web - WWW '16, pp 287–297. https://doi.org/10.1145/2872427.2883041 http://arxiv.org/abs/1602.00370

175. Tenenbaum JB, De Silva V, Langford JC (2000) A global geometric framework for nonlinear dimensionality reduction. Science 290(5500):2319–2323

176. Tipping ME, Bishop CM (1999) Probabilistic principal component analysis. J Roy Statist Soc Ser B (Statist Methodol) 61(3):611–622. https://doi.org/10.1111/1467-9868.00196. https://rss.onlinelibrary.wiley.com/doi/abs/10.1111/1467-9868.00196

177. Torgerson WS (1952) Multidimensional scaling: I. Theory and method. Psychometrika 17(4):401–419

178. Tukey JW, Tukey PA (1988) Computer graphics and exploratory data analysis: an introduction. In: Collected works of John W Tukey: Graphics: 1965–1985, vol 5, p 419

179. Udell M, Horn C, Zadeh R, Boyd S (2016) Generalized low rank models. Found Trends® Mach Learn 9(1):1–118. https://doi.org/10.1561/2200000055. https://www.nowpublishers.com/article/Details/MAL-055

180. Uhlmann JK (1991) Satisfying general proximity / similarity queries with metric trees. Inf Process Lett 40(4):175–179. https://doi.org/10.1016/0020-0190(91)90074-R. http://www.sciencedirect.com/science/article/pii/002001909190074R

181. Van Der Maaten L (2009) Learning a parametric embedding by preserving local structure. In: Artificial intelligence and statistics, pp 384–391

182. Van Der Maaten L (2014) Accelerating t-SNE using tree-based algorithms. J Mach Learn Res 15(1):3221–3245

183. van der Maaten L, Hinton G (2008) Visualizing data using t-SNE. J Mach Learn Res 9(Nov):2579–2605

184. Van Der Maaten L, Postma E, Van den Herik J (2009) Dimensionality reduction: a comparative. J Mach Learn Res 10(66-71):13

185. Venna J (2007) Dimensionality reduction for visual exploration of similarity structures. PhD thesis, Helsinki University of Technology, Espoo, oCLC: 231147068

186. Venna J, Kaski S (2001) Neighborhood preservation in nonlinear projection methods: an experimental study. In: International conference on artificial neural networks. Springer, Berlin, pp 485–491

187. Venna J, Kaski S (2006) Local multidimensional scaling. Neural Netw 19(6–7):889–899. https://doi.org/10.1016/j.neunet.2006.05.014. http://linkinghub.elsevier.com/retrieve/pii/S0893608006000724

188. Venna J, Kaski S (2007) Nonlinear dimensionality reduction as information retrieval. In: Artificial intelligence and statistics, pp 572–579

189. Venna J, Peltonen J, Nybo K, Aidos H, Kaski S (2010) Information retrieval perspective to nonlinear dimensionality reduction for data visualization. J Mach Learn Res 11(Feb):451–490

190. Vidal Ruiz E, Casacuberta Nolla F, Rulot Segovia H (1985) Is the DTW "distance" really a metric? an algorithm reducing the number of DTW comparisons in isolated word recognition. Speech Commun 4(4):333–344. https://doi.org/10.1016/0167-6393(85)90058-5. http://www.sciencedirect.com/science/article/pii/0167639385900585

191. Villalva MG, Gazoli JR, Filho ER (2009) Comprehensive approach to modeling and simulation of photovoltaic arrays. IEEE Trans Power Electron 24(5):1198–1208. https://doi.org/10.1109/TPEL.2009.2013862

192. Vlachos M, Domeniconi C, Gunopulos D, Kollios G, Koudas N (2002) Non-linear dimensionality reduction techniques for classification and visualization. In: Proceedings of the eighth ACM SIGKDD international conference on knowledge discovery and data mining, pp 645–651

193. Vladymyrov M (2019) No pressure! addressing the problem of local minima in manifold learning algorithms. In: Wallach H, Larochelle H, Beygelzimer A, Alché-Buc Fd, Fox E, Garnett R (eds) Advances in neural information processing systems, vol 32. Curran Associates, Red Hook, pp 680–689. http://papers.nips.cc/paper/8357-no-pressure-addressing-the-problem-of-local-minima-in-manifold-learning-algorithms.pdf

194. Vladymyrov M, Carreira-Perpinán MÁ (2013) Entropic affinities: properties and efficient numerical computation. In: Proceedings of the 30th international conference on machine learning, no 3, pp 477–485

195. Wang Baldonado MQ, Woodruff A, Kuchinsky A (2000) Guidelines for using multiple views in information visualization. In: Proceedings of the working conference on advanced visual interfaces, association for computing machinery, New York, NY, AVI '00, p 110–119. https://doi.org/10.1145/345513.345271

196. Ward Jr JH (1963) Hierarchical grouping to optimize an objective function. J Am Statist Assoc 58:236–244. https://amstat.tandfonline.com/doi/pdf/10.1080/01621459.1963.10500845?needAccess=true#.XvHBOWgzaUl

197. Watson GS (1964) Smooth regression analysis. Sankhyā Indian J Stat Ser A (1961–2002) 26(4):359–372. https://www.jstor.org/stable/25049340

198. Welch P (1967) The use of fast Fourier transform for the estimation of power spectra: a method based on time averaging over short, modified periodograms. IEEE Trans Audio Electr 15(2):70–73. https://doi.org/10.1109/TAU.1967.1161901

199. Wertheimer M (1923) Untersuchungen zur Lehre von der Gestalt. II. Psycholog Forschung 4(1):301–350. https://doi.org/10.1007/BF00410640

200. Wilkinson L, Anand A, Grossman R (2005) Graph-theoretic scagnostics. In: IEEE symposium on information visualization, 2005. INFOVIS 2005. IEEE, Piscataway, pp 157–164

201. Wismüller A, Verleysen M, Aupetit M, Lee JA (2010) Recent advances in nonlinear dimensionality reduction, manifold and topological learning. In: ESANN, European symposium on artificial neural networks

202. Yang Z, King I, Xu Z, Oja E (2009) Heavy-tailed symmetric stochastic neighbor embedding. In: Advances in neural information processing systems, pp 2169–2177

203. Yang Z, Peltonen J, Kaski S (2013) Scalable optimization of neighbor embedding for visualization. In: International conference on machine learning, pp 127–135

204. Yang Z, Peltonen J, Kaski S (2014) Optimization equivalence of divergences improves neighbor embedding. In: International conference on machine learning, pp 460–468

205. Yianilos PN (1993) Data structures and algorithms for nearest neighbor search in general metric spaces. In: Proceedings of the fourth annual ACM-SIAM symposium on discrete algorithms, pp 311–321

206. Young G, Householder AS (1938) Discussion of a set of points in terms of their mutual distances. Psychometrika 3(1):19–22

207. Zhang Sq (2009) Enhanced supervised locally linear embedding. Pattern Recog Lett 30(13):1208–1218. https://doi.org/10.1016/j.patrec.2009.05.011. http://www.sciencedirect.com/science/article/pii/S0167865509001202

208. Zhao L, Zhang Z (2009) Supervised locally linear embedding with probability-based distance for classification. Comput Math Appl 57(6):919–926. https://doi.org/10.1016/j.camwa.2008. 10.055. http://www.sciencedirect.com/science/article/pii/S0898122108005695

209. Zheng JX, Pawar S, Goodman DFM (2019) Graph drawing by stochastic gradient descent. IEEE Trans Vis Comput Graph 25(9):2738–2748. https://doi.org/10.1109/TVCG.2018. 2859997

210. Zhou H, Yuan X, Qu H, Cui W, Chen B (2008) Visual clustering in parallel coordinates. Comput Graph Forum 27(3):1047–1054. https://doi.org/10.1111/j.1467-8659.2008.01241. x. https://onlinelibrary.wiley.com/doi/10.1111/j.1467-8659.2008.01241.x

211. Zou H, Hastie T, Tibshirani R (2006) Sparse principal component analysis. J Computat Graph Statist 15(2):265–286

212. Zwan Mvd, Codreanu V, Telea A (2016) CUBu: universal real-time bundling for large graphs. IEEE Trans Vis Comput Graph 22(12):2550–2563. https://doi.org/10.1109/TVCG. 2016.2515611

Index

Printed in the United States
by Baker & Taylor Publisher Services